ENCYCLOPEDIA OF ANTIQUE

SCIENTIFIC INSTRUMENTS

ENCYCLOPEDIA OF ANTIQUE
SCIENTIFIC
INSTRUMENTS

by John FitzMaurice Mills

Facts On File Publications
460 Park Avenue South
New York, N.Y. 10016

Published in the United States of America in 1983 by Facts On File, Inc.,
460 Park Avenue South, New York, NY 10016
Published in Great Britain in 1983 by Aurum Press.

Library of Congress Cataloging in Publication Data
Mills, John FitzMaurice, 1917 –
Encyclopedia of antique scientific instruments.
Bibliography: p.
1. Scientific apparatus and instruments – Dictionaries.
2. Scientific apparatus and instruments – Collectors and
collecting. I. Title.
Q184.5.M54 1984   681'.75'0321   83-5676
ISBN 0-87196-799-5
Printed in Great Britain
First printing

CONTENTS

PICTURE ACKNOWLEDGEMENTS

The author and publishers wish to thank the following for their permission to reproduce photographs.
Black and white Christie's: pp. 56, 57, 59, 61, 65, 82, 85 (*left, right*), 86 (*left, right*), 87, 103, 113, 126, 142, 155 (*right*), 159, 161, 180, 184 (*bottom*), 193, 194, 202 (*top*), 213, 221 (*left*), 229. Mary Evans Picture Library: 49, 52, 72, 94, 107, 108, 121, 123, 135, 137, 140. Phillips: 48, 51, 60, 63, 64 (*top, bottom*), 66, 74, 75, 80, 81, 102, 106, 112, 125 (*bottom*), 136, 145, 152, 153, 154, 155 (*left*), 156, 164, 167, 169, 173, 181, 184 (*top*), 189, 190, 191 (*left, right*), 195, 198 (*left, right*), 200, 208, 211, 216, 219, 220 (*top, bottom*), 221 (*right*). St Patrick's College Museum, Maynooth: 76, 83, 99, 125 (*top*), 150, 176, 209, 222. Sotheby, Parke Bernet & Co.: 68, 89, 95, 101, 104, 110, 111, 139, 144, 166, 179, 202 (*bottom*), 204 (*left, right*), 205, 212, 227.
Colour Cavendish Gallery: V. Christie's I, IV (*bottom*), VI (*top, bottom*), VIII, IX (*top*). Phillips: II (*top, bottom*), IV (*top*), VII, IX (*bottom*), X-XVI.

AUTHOR'S NOTE

I have chosen to include among the entries certain astronomical, geological and other scientific terms and inventions that have particular bearing upon the instruments described or are associated with their makers.

INTRODUCTION

While there is a difference between scientific instruments and other works of art and craftsmanship, these objects are often products of genius with a direct appeal to our aesthetic sensibility. Fine furniture, silver, paintings, and sculpture may be more traditional subjects of cultural appreciation, but contemporary interest in the history of science has given a new impetus to the appreciation of those instruments associated with scientific progress. They have an extra fascination for the collector.

Today the majority of antique scientific instruments may be viewed in museums, yet there is still considerable scope for the collector. There are a number of dealers who confine their activities solely to this field; the great salerooms have specialist dates and, increasingly, both local and international antique fairs are mounting displays to draw the attention of their visitors. There are various reasons for this. Compared with the world of centuries ago, or even fifty years ago, life is dominated by science. Its enveloping influence has awakened in many the exploratory spirit that wants to understand more of the ways and means of things. But, most of all, there is in the examination and handling of some instrument from the past a communion with the original inventor and user, a sensation of being at least in the wings of the discovery that this telescope, microscope, or navigational aid made possible. With this is an appreciation of the craftsmanship, design and use of materials.

There seems little to connect science and art today, but in earlier times many of the instruments now sought after by collectors were made at the behest of a patron who was interested not only in the sciences but also in the arts. Thus there is an attention to design and applied decoration that is absent from the clinical functionalism of today's instruments. The historian can draw closer to the atmosphere of the time he is studying by examining and using the tools of those who first set out on a journey of exploration to the heavens, in micrology or some other field. The 20th-century designer of instruments can still gain much by going back to look again at how the basic principles were discovered and adapted by the pioneers.

Even a brief study of the history of these instruments can be sufficient to open the mind to the wonders of scientific progress. From the start it appears as a never-ending search, with minds leap-frogging over each other in an exploration that seems to become more and more vast with each step that is taken. One telescope will bring into view a hundred new stars, the next will multiply this result many times over. Each fresh uncovering by the enlightened mind of a genius points to the almost unfathomable dimensions and infinite possibilities of what lies ahead.

It is a strange and sad comment on our social structure that very often the achievements of these pioneers bring down on them a totally undeserved attack. The 'establishment' of practically every period in history appears to have been loath to accept and appreciate the new, whether in the sciences, the arts or any other sphere. One of the clearest instances of this is the treatment meted out to Galileo (*q.v.*), when he was forced into theological controversy. Most scientists who are bent on breaking through the obscuring clouds of ignorance are men of honesty, having no time for intrigue. It is very often this quality that seems to be their undoing.

Whereas the artist at work on a painting is usually conscious that he is endeavouring to bring into being a work of art, or the potter modelling some small exquisite figure is involved in the creation of a beautiful form, the maker of instruments, although he may have the aesthetic eye of a trained designer or artist, is first or foremost creating tools to achieve some scientific result. It may be the simplest gnomon of an early Egyptian, trying to divide the light of day into a system of workable divisions, a heavy bronze navigational aid, an orrery or an early thermometer. In many cases the formative story behind the instrument is a long one of trial and error. There may have been flashes of invention, but development is well summed up by Thomas Edison with the words: 'One per cent inspiration – 99 per cent perspiration.' His own struggle to produce perhaps one of the most useful of instruments – the electric light bulb – is evidence enough. In addition to the thousands of experiments he made, he filled some 200 notebooks with 40,000 entries. Edison, who also numbered the gramophone among his inventions, must hold the record for patents taken out to protect his discoveries – 1,693 in all.

An individual cannot hope to collect from the whole field of scientific instrumentation. To cover the ground at all comprehensively, the resources of a major museum are needed. What, then, are the principal categories to investigate? For many, the primary attraction will probably be astronomy, navigation or some form of measurement, but others are interested in microscopes, timekeepers, pharmaceutical equipment, physical apparatus, medical and dental instruments, cameras and projectors and even delicate machinery or unusual experimental devices, of which there have been some truly fascinating objects.

Around 1800 there was made the so-called 'Temple of Vesta' which was an example of a chemical instant fire machine. Vesta was the Roman goddess associated with the hearth, and a flame was kept burning and jealously guarded in her temple by the Vestal Virgins. The small temple 'machine' was a work of art in its own right, designed with architectural elegance. Inside, hydrogen was generated and, by the pressing of a button, released through the mouth of a miniature lion and at the same time ignited by a small spark from a concealed electrostatic device.

Early in the 18th century, it became the fashion to commission strange instruments and devices to show scientific principles in a way that would entertain as well as educate. One of the leading figures who collected and displayed these extravagances was the Rev. J. T. Desaguliers (1683–1744), who brought them together for the edification of British Royal children. The collection was added to until about the end of the 18th century and is today known as the George III collection. The range encompassed is wide, both historically and in the scope of the sciences that it touches. There is a model of the hydraulic screw for raising water, for which Archimedes (q.v.) is given credit; this was made by George Adams (q.v.) who also made the exhausting pump, which had improvements by Smeaton (q.v.). This type of pump seems to have been an essential item for the specialist collector of the time. It could be used to demonstrate the phenomena of objects falling and the non-transmission of sound. In a more macabre vein, our ancestors apparently experimented with the effects of reduced and increased atmospheric pressures on living creatures. Thomas Hornsby (1763–1810), reader in experimental philosophy at Oxford, when beginning to draw the air out of a glass chamber in which was contained an unfortunate cat, would remark 'You will observe gentlemen, that the animal exhibits symptoms of uneasiness ... The animal seems to be considerably incommoded ...' It is recorded that once he became so immersed in his research that a certain lady's pet was only saved from extinction by the timely intervention of his servant.

It may be seldom today that these devices of fantasy are found outside a museum or a long-established private collection, yet the possibility still exists. If they come up in the salerooms it is almost certain that they will be heavily competed for. Even the examples of what might be termed 'scientific archaeology' send the bids rising, as evidence the three figure sums that old photographic prints command. Specialist dealers will have fine instruments but they will also be asking relative market prices.

The tyro collector can try, as a possible resort, hours of jumbling through junkshop piles and battered cardboard boxes. There is the chance that some sadly mutilated specimen may emerge, or some parts of a device that has been dismembered in ignorance – lenses from a telescope, fragments of a compass, a dial or arm from a nocturnal. Old house contents' sales can sometimes produce the unexpected; whilst the auctioneer is working his way down through the furniture, pictures, silver and the rest, one can retire and finger through those last, pitiful lots. Here, a number of indescribable items are often lumped together. A damp part-rotten old wooden box may contain an early beam compass, which, to the uninitiated eye, may appear as a bundle of tarnished brass rods. A bent tube with a few fragments of decorated leather still adhering to it might be the barrel of a Culpeper microscope. A heavily corroded green patinated bronze ring could be the remains of a mariner's astrolabe. So much has been allowed to depreciate through ignorance or, almost worse, because the collection and care of scientific instruments was not fashionable.

To look at and, if possible, touch some of these early instruments brings one into a closer association with more than just the materials from which they are made. It seems that the atmosphere which pervaded the original challenge and solving of a problem is recreated.

The early globe which revolves on its stand, with its very imperfect rendering of the layout of the countries of the world, may appear almost comic by the knowledge of today; yet it is a great deal more beautiful than our contemporary examples. Although the makers and the cartographers may have been a long way out, it must be remembered that they had, at least, made one very great contribution: they were admitting they no longer believed the earth to be a great flat disc.

A glance at some salt-corroded old marine navigation instrument can set the imagination running. Was this used by a captain who sailed down the Channel with Drake, or was it in the hands of a freebooting adventurer who was trusting his life and the success of his venture to this small scientific thing of bronze, brass or iron?

Yet the single instrument which has opened up the widest horizons and holds the greatest fascination for many collectors is the telescope. Prior to the work of Galileo and his contemporaries the observation of the heavens had been solely by the eye; movements of the celestial bodies could be noted, although they remained distant sparks of light in the night skies. Then a tube and two lenses magically brought these sparks nearer. From being mere pinpoints, some of them could be seen as spheres and more than the surface could be observed.

Nevertheless, this wider outlook could only be achieved once philosophical and theological dogma had been swept away. Such dogma prevailed from *c.*200 BC onwards, when the splendid achievements of Aristarchus and others could be decried by Plato and other philosophers, who demanded that the earth must fit their theories. It could well be that without such obstacles, the work of Kepler (*q.v.*) would have been anticipated by many centuries.

ASTRONOMY

Early agricultural man had no calendars or clocks to tell him when to sow or reap or hunt so he turned his eyes heavenwards and learned to read the movement and placings of the stars, sun and moon. Slowly he appreciated that a recurrent order of celestial appearances could tell him when to sow and when to expect the harvest.

In China about 2300 BC, the emperor Yao issued orders fixing the equinoxes and solstices by this same movement of the stars. These orders were recorded in the *Shu Chung*, a collection of antique documents of the time of Confucius (550–478 BC). It is known that Yao was not the first to use such a system; it had already been long established. It is interesting to note, in another part of the *Shu Chung*, the sad fate of the official astronomers Hsi and Ho who failed to perform the correct rites during an eclipse of the sun and were executed. Later findings have established that, close to Yao's time, there was a partial eclipse which could have been seen in northern China in the year 2136 BC.

In the first millennium BC, the Chinese were making observations on the meridian; they were using water clocks and possessed measuring instruments similar to armillary spheres and quadrants. About 1100 BC Chou Kung, a learned mathematician, worked out with great accuracy the obliquity of the ecliptic, although at the same time he failed completely when he tried to work out the distance of the sun from the earth; this was largely because he accepted as fact that the earth was a giant flat disc and not spherical. There are records of observations of the behaviour of comets dating from 2296 BC. The earlier ones may be inaccurate, but modern findings have pointed to the fact that those from 611 BC onwards are correct.

There were two instruments that were made at the time of Kublai Khan (*c*.1280) which up to the end of the 19th century were still preserved in Peking. These were provided with large graduated circles adapted for the measurement of declination and right ascension and proved that at least three centuries before Tycho Brahe (*q.v.*), the Chinese had versions of his most important inventions. The native progress and theories of astronomy in China were superseded in the 17th century by the more advanced scientific teachings from Europe, taken into that country by the Jesuit missionaries. However, the Chinese globe of *c*.1627 in the British Museum, which bears an inscription saying that the earth is the centre of the universe and is bolt upright, that is, that there is no obliquity of the ecliptic, was made under Jesuit instruction.

In the Middle East, particularly in Egypt, astrolatry was a curtain raiser to the study of astronomy. The stars were seen as objects of worship. At first, their rising was largely linked to ritual observance and only later was it understood that there could be a relationship to acts of husbandry and other aspects of civil life. The length of the year was accurately fixed against the recurring floods of the Nile. There is an example of the curiously accurate orientation of the pyramids that points to a highly developed skill in the watching of the heavens attained by the early Egyptians of the third millennium BC.

The Greeks derived many of their basic theories on astronomy from the Babylonians. They copied the Babylonian asterisms and learnt from them how to predict eclipses by means of the 'Saros', discovered at an unknown time in Chaldea. This is a cycle of 18 years and 11 days, or 223 lunations, at the end of which the moon very nearly returns to her original position with regard to her own nodes and perigee, and also to the sun. It is not possible to point to the date that the study of astronomy along the banks of the Euphrates began, but there are records dating from the reign of Sargon of Akkad (3800 BC) which imply that the various aspects of the heavenly bodies had been under observation and consideration for a long time. At this time astronomy leaned heavily on legend and the signs and meanings of the zodiac, but gradually Babylonian astronomy purified itself from astrological influence and was used as an adjunct to regulate many facets of civil life. The Jesuit fathers, Epping and Strassmeier, deciphered and interpreted a number of clay tablets in the British Museum, which give detailed information of techniques used in Mesopotamia during the 2nd century BC. These show no traces of Greek influence and really are a sublimation of a long local tradition. The Babylonians had worked out revolutionary cycles for the planets, and knew that the planet Venus returns in exactly eight years to a given starting point in the sky. Further, they had found out similar periodic relations of 46, 59, 79 and 83 years for Mercury, Saturn, Mars and Jupiter respectively.

From the 7th century BC there was a constant flow of knowledge from East to West. In 640 BC a Babylonian scholar, Berossus, founded a school on the island of Cos and it is possible that amongst his pupils may have been Thales of Miletus (q.v.). Pythagoras of Samos (q.v.), picked up during his travels in Egypt and the Middle East the ability to identify the morning and the evening stars, to see and understand the obliquity of the ecliptic. More than this, he saw the earth as a sphere, poised in space as a free body.

The first person to develop a mathematical theory of celestial appearances was Eudoxus of Cnidus (q.v.). He tried to solve the problem of producing combined uniform circular movements to correspond to the resultant effects actually seen. To this end he gave the sun and moon and the five planets a set of variously revolving spheres to the total number of 27. The 'Eudoxian' or 'homocentric' system was further elaborated by Callipus and Aristotle (*q.v.*) and modified by Apollonius of Perga (250–220 BC) into the hypothesis of deferents and epicycles which, for 1,800 years, embodied the characteristic ideas of Greek astronomy. The peak of Greek achievement was reached in the school of Alexandria. Shortly after its foundation the two scholars Aristyllus and Timocharis (*c.*320–260 BC) produced the first catalogue to give star positions, measured from a reference point in the heavens. Aristarchus of Samos (*q.v.*), working at Alexandria, produced a treatise on the magnitudes and distances of the sun and moon which in the late 17th century was edited by John Wallis. It gave a theoretically sound method for determining the relative distances of sun and moon by measuring the angle between their centres when half the lunar disc was illuminated. Unfortunately the time of dichotomy was varied and useful results could not be obtained. Ptolemy Euergetes brought Eratosthenes (*q.v.*), a Cyrenean, to Alexandria to take charge of the royal library. But whilst there the scholar invented or improved armillary spheres, the chief instruments of ancient astronomy.

From antiquity, two figures stand out as astronomers of skill and importance, above most of the others. These are Hipparchus (*q.v.*) and Ptolemy (*q.v.*). Both men, to a large degree, thought along similar lines and when the personal work of each is compared it is evident that in their time they transformed celestial science. Hipparchus determined the chief points of astronomy, the lengths of the tropical and sidereal years, of the months and of the synodic periods of the planets. He constructed armillary spheres with sights for making observations. It is to his credit that he established astronomy on a firm geometrical basis. Ptolemy's system was the admiration of the savants of the time for the great ingenuity he employed to bring it to perfection. He made a globe which included the motions of the sun and moon, and described this in his book, the *Almagest*, which was the sublimation of Greek astronomy.

With the sacking of Alexandria by Omar in AD 641 the spirit of scientific experiment in the area faded. In about 800 Harun al-Rashid ordered the first Arabic translation of the *Almagest*. This was followed by another, commissioned by the Caliph al-Mamun, who in 829 built the great Observatory at Baghdad. Here it was that Albumazar (805–85) watched the skies and set his horoscopes. Here also Tobit ben Korra (836–901) worked out his strange and, for a long time, misleading theory of the 'trepidation' of the equinoxes. Though not working in Baghdad, Ibn Junis (*c.*950–1008), used the theories of the scholars from there. He compiled the Hakimite Tables of the planets. When in Cairo, in 977 and 978, he observed two solar eclipses, which were the first to be recorded with scientific accuracy.

In Persia, Hulagu Khan, who died in 1265, established an Observatory at Maraga and put it under the directorship of Nasir ud-din (1201–74), who drew up the Ilkhanic Tables. The Observatory at Maraga was equipped, amongst other things, with a mural quadrant of 12 ft radius and altitude and azimuth instruments. To the north Ulugh Beg (*q.v.*), the grandson of Tamerlane, built a magnificent Observatory at Samarkand in 1420, where he studied and checked nearly all Ptolemy's stars, and the tables he published were in constant use for nearly two centuries afterwards.

The Moors introduced Arab astronomy to Spain and for a time its study flourished at Cordoba and Toledo. From the latter city came in 1080 the Toletan Tables which were worked out by Arzachel. Later the Alfonsine Tables, which were published in 1252, were prepared under the patronage of Alfonso X of Castile. The appearance of these tables marked the beginning of European science and they came out very close to the date of a textbook on spherical astronomy, *Sphaera Mundi* by John Holywood, a Yorkshire man, which was very popular and eventually went to 59 editions. In the 15th century German scientists worked to improve on the Ptolemaic theories. A leader in the field of astronomy was George Purbach (1423–61) who introduced to Europe the findings on altitude determination that had been worked through by Ibn Junis. Purbach was to have considerable influence on the Nuremberg scholar Bernard Walther (1430–1504) who had an observatory which contained clocks driven by weights and had made several breakthroughs in the field of practical observation. Walther provided the astronomer Regiomontanus (*q.v.*) with instruments which would today be worth a fortune. When Walther died his heirs gave away his personal collection. Old pieces of brass (as they thought) went to the rag-man and his manuscripts were thrown away as unwanted rubbish.

The latter part of the 15th century and the first half of the 16th were to see a dramatic reform of ideas from scholars bypassing the Ptolemaic theories. One man who, more than any other, caused this revolution was Nicolaus Copernicus (*q.v.*). In 1543 he published *De Revolutionibus Orbium Coelestium*. Sadly for him, the great moment of publication was marred by an uninvited foreword to the book, which claimed that the matter set out by him was hypothesis and not definitely established fact. Nevertheless, Copernicus had completed the labour of almost turning astronomy inside out. In his new system, the sphere of the fixed stars no longer revolved diurnally; instead the earth was seen as rotating on an axis directed towards the celestial pole. The sun was stationary, whilst the planets, including the earth, revolved round the sun. Copernicus strangely enough retained the old idea of a circular motion, instead of an elliptical movement, for these bodies.

In 1551 Erasmus Reinhold (1511–53) published *Tabulae Prutenicae*, which was calculated on the Copernican principles, and although these were an advance on the Alfonsine Tables, there were still areas of doubt. Almost simultaneously two people set about to reform the methods. They were the Landgrave William IV of Hesse-Cassel (1532–92) and Tycho Brahe (*q.v.*). At Cassel in 1561 the Landgrave built the first observatory to be fitted with a revolving dome, and he worked on a star catalogue, which actually was never completed. For assistants he had Christoph Rothmann and Justus Byrgius (*q.v.*). The latter was particularly skilful at time determinations for measuring right ascensions and was also the maker of exquisite mathematical instruments. It is believed that the altitude and azimuth instrument made its first appearance in Europe at Cassel.

Tycho Brahe can be given the credit for bringing pre-telescopic observation to its highest peak. His instruments were often large and of a type that had not been used since the days of Nasir ud-din. In 1569 at Augsburg he constructed a 19 ft quadrant and a celestial globe 5 ft in diameter. After Brahe's death, his son-in-law Tengnagel, having failed to sell the instruments that played such a part in the development of astronomy, left them to rot to such an extent that they were to become scrap metal in a few years. Brahe's work and notes and discoveries were inherited by the German astronomer Johann Kepler (*q.v.*), who was born at Weil in the Duchy of Württemberg. Kepler's main contributions to the growing story of astronomy were perfecting the geometrical plane of the solar system and shedding light on the predicting power of astronomy. In 1627 the Rudolphine Tables were published; these were prepared by Kepler from elliptic elements, and for over a century were considered the leading authority for reference and, in principle, have still not been superseded.

Before the 17th century a telescope, that is, an instrument which employs two or more lenses, had not been invented although it may have been that someone noticed that the use of a single lens would magnify objects at a distance. This will occur if the lens is held at almost arm's length from the eye; it is a simple experiment that can be tried with a pair of reading or other glasses. However, at the beginning of the 17th century there happened one of those odd events which have often affected progress and growth in history. This time it was the discovery of a telescope, credit for which is given to a spectacle maker, Hans Lippershey (*q.v.*) of Middleburg on the island of Walcheren. The story has it that in 1608 he was holding two lenses some distance apart, happened to see the neighbouring church spire through them and noted how much larger it appeared. He promptly mounted two similar lenses in a tube and made what in all likelihood was the world's first telescope;

this he then proudly presented to the States-General Prince Maurice. As with such happenings the news spread quickly across Europe and it is said was picked up by Galileo in May 1609 in Venice.

Galileo Galilei, from his early childhood, appeared as a prodigy in many fields. Amongst his favourite pursuits when still quite young was designing and making ingenious mechanical toys. The news which he received about the discovery by the spectacle maker in May 1609 must have seemed like a key he had been subconsciously looking for. It is recorded that after only a day's study of Lippershey's discovery, Galileo was able to invent his telescope. His first instrument had a three-fold magnifying power but he rapidly improved upon this until he had attained a power of thirty-two. He made his instruments with his own hands; in fact, he manufactured hundreds which were soon in demand by astronomers from all over Europe. The telescope that Lippershey invented was like our opera glasses, that is, it had a convex and a concave lens. Kepler was to suggest a better instrument with two convex lenses. Although this would produce an upside down image in observing the stars, it was not of great consequence. The insertion of a third convex lens would bring the image the right way up.

When Galileo pointed his instrument towards the heavens a fresh and exciting period of great exploration was opened up. His discoveries and his telescopes were described in his book *Sidereus Nuncius*, which was published in Venice in 1610. For the first time a believable description of the surface of the moon was given and the phases of the planet Venus and the four satellites of Jupiter were described, all of which lent support to the theories of Copernicus. Galileo also observed sunspots and recorded that they changed position from day to day, which showed to him that the sun revolved upon an axis. Galileo unquestionably did more than any previous astronomer to pass on his knowledge in a manner that could be easily understood by the general public. His inventive mind encompassed much and, besides numerous theories and discoveries, he gave the world the compass and thermometer.

A contemporary of Galileo was a talented amateur astronomer Johann Bayer (1572–1625), born at Augsburg. In 1603 he produced his *Uranometria* which contained 51 plates based on designs by Albrecht Dürer and was the first publication to contain charts showing some of the southern constellations.

Sir Isaac Newton (*q.v.*) was born in Colsterworth in Lincolnshire. He displayed a genius for mechanics at an early age and made water clocks, kites and other devices; he was also said to have invented a strange four-wheeled carriage which could be moved by the rider. He entered Trinity College, Cambridge, where one of the early influences on him was Descartes's

Geometry. As with Galileo, Newton's talents were diverse, but his principal achievements were first in the identification by strict numerical comparisons of terrestrial gravity with the mutual attraction of the heavenly bodies, and, secondly, following this fact's mechanical consequences throughout the solar system. Despite the immensity of his discoveries and theories, Newton must have had, like so many great personalities, a disarming humility, for he once said about himself: 'I do not know what I may appear to the world but to myself I seem to have been only like a boy, playing on the seashore, and diverting myself in now and then finding a smoother pebble or a prettier shell than ordinary, whilst the whole ocean of truth lay all undiscovered before me.'

It would probably have brought quiet pleasure to him if he could have known that designs of ships and aircraft would result from his discussions in Book II of the *Principia*, where he looked into the movement of bodies through a 'resisting medium' like water or air, and the launching of space probes would be based on his theory of gravitation.

According to the French astronomer Lagrange, 'Newton was the greatest genius that ever existed and the most fortunate, for we cannot find more than once a system of the world to establish.'

Newton was the first person to make a reflecting telescope, which he produced in 1668. The basic idea had actually been suggested by James Gregory (*q.v.*) in 1663 but had not been developed. Newton turned his attention towards the idea of a reflecting telescope after his experiments on the refraction and dispersion of light through glass prisms. He concluded that lenses would always produce chromatic aberration, which meant that the images would always have coloured edges. In 1671 Newton made a second reflecting telescope which is still preserved at the Royal Society.

Up to the middle of the 19th century, mirrors were made of metal, usually alloys. These could be given a high polish but in damp atmospheres could quite quickly become dull and needed constant attention. Newton, for his mirrors, used an alloy known as bell-metal, which was composed of two parts tin and six parts copper. The early development of the telescope was greatly hampered by the poor quality of the glass available to the lens maker. It was almost impossible to grind a lens of more than about 4 in diameter, as the raw glass contained many defects. This did not greatly trouble the makers of microscopes and spectacles, but it held back the telescope designers.

The history of glass making probably started in ancient Egypt. Here the basic materials were readily available and the potters were skilled in the production of vitreous glazes. There is a wall painting at Beni Hasan dating from the Eleventh Dynasty, which could be taken as showing an accurate picture of glass blowing. This reading of the subject was disputed by Dr

Flinders Petrie, who felt it only showed some metallurgical process, with reeds dipped into clay. It was Dr Petrie who, at Tell el Amarna, found what can be claimed to be the earliest examples of Egyptian glass known, and these belong to the Eighteenth Dynasty.

The most rapid development in the manufacture of glass was achieved by the Romans, who produced glass of a very high quality and used it for a wide number of purposes. Before their civilization was crushed by the Hunnish vandals, the Romans had begun to glaze windows and to make primitive optical instruments.

From then the progress towards a fine glass was taken up in a number of centres, notably Venice from about the 6th century and later in France, Britain and Germany. In Germany, in 1787, Joseph von Fraunhofer (q.v.) was born, the son of a glazier. He was apprenticed to Weichselberger, a glass polisher and a looking-glass maker. Fraunhofer's story had a strange start, as apparently on 21st July, 1801 he nearly lost his life when the house in which he was lodging collapsed. His extrication from the ruins was witnessed by Maximilian Joseph, the elector of Bavaria, who rewarded his escape with 18 ducats. Fraunhofer was not only able to buy himself out of his apprenticeship to a glass polisher but also to purchase a glass-polishing machine. From then he began to study optics and to make optical glasses. In 1806 he was appointed as optician to the mathematical institute which had been founded in Munich by Joseph von Utzschneider. Here the founder arranged for him to be instructed in the making of flint and crown glass by Pierre Louis Guinand.

In 1809 Fraunhofer, with Utzschneider and Reichenbach, set up an optical institute at Benedictbeuren near Munich. This was moved into the city in 1818 and after Fraunhofer's death was managed by G. Merz. Amongst the developments made by Fraunhofer was a machine for polishing mathematically uniform spherical surfaces. He was also the maker of the great telescope at Dorpat which had an objective lens with a diameter of $9\frac{1}{2}$ in. As an illustration of his advanced skill there is a letter to Sir William Brewster, in which the optician offered to make an achromatic glass with a diameter of 18 in.

The design of telescopes is a fascinating story. Take, for example, the huge 150 ft object built by Hevelius of Danzig, the giant Hale telescope of Mount Palomar in South California with a 200 in reflecting disc, and the vast reflector in the Caucasian mountains with a mirror of 6 metres diameter.

Telescope makers became as famous as clock makers and specimens of their work, even in their own times, were eagerly sought after. One such maker was Christopher Lock, working in London in 1673. He was one of the leading optical glass grinders of his time and was commissioned by the Royal

Society to make various telescopes, including a larger copy of Newton's reflector. Very long telescopes, such as Hevelius's *150 ft object, are known as aerial telescopes; these allowed for observation but were not suitable for making angular measurements. Huygens* , using an aerial telescope with a focal length of 210 ft, in 1655 discovered the first satellite of Saturn. The possibilities were widened with the theory of James Gregory of Aberdeen in his book *Optica Promota*, which was published in 1663, where he set out the idea for the first reflecting telescope with two concave mirrors. The collector would be paraboloidal and the small reflector ellipsoidal, which would focus the image of the speculum. In the centre of this would be a small aperture which would allow the light to pass to the eyepiece. Gregory himself had two London opticians make the necessary reflectors, but their work did not match up to the quality he was seeking and it appears that he lost interest in the project. The theories were not lost, however, because John Hadley (*q.v.*) in 1726 did produce a Gregorian telescope which was successful.

Other instrument makers, encouraged by Hadley's success, turned to the production of similar designs. One who was skilled was George Adams (*q.v.*), who was appointed instrument maker to the Prince of Wales before he became George III in 1760, and who continued in this position until his death in 1772 when his son, also George, took over the appointment. One of the most brilliant telescope makers of his time was James Short, who was responsible for developing speculum mirrors to a very high degree of accuracy, using methods which were kept secret and have never been published.

William Herschel (*q.v.*) was born in Germany in 1738 but came to England at the age of 19. He took up astronomy at first as a spare time interest but he soon became fascinated with instruments, particularly telescopes. Between 1774 and 1789 he built a large number of specula of gradually increasing size, up to 4 ft in diameter. These were ground with great skill and care and their final optical excellence was proved by a number of exciting discoveries. In 1781 Uranus was recognized by its disc; in 1787 Oberon and Titania, two of its satellites, were seen. Two years later he saw Mimas and Enceladus, the innermost moons of Saturn. Despite the importance of these discoveries, they were not really Herschel's main goals. He had set himself to explore the sidereal heavens.

Another astronomer, Charles Messier, in 1781 had catalogued 103 nebulae. This was far surpassed by Herschel who raised the total to 2,500, classified the nebulae, worked out the laws of their distribution and pointed to their place in a scheme of development. In 1783 he established the direction of the sun's movement in space and from this was able to work out a plan of order for the stellar system. As his fame grew, William Herschel was appointed

Astronomer Royal to George III in 1782 and moved from Bath to Datchet in that year and to Slough four years later, where he remained until he died in 1822. His son carried on his father's work and extended the observation to include not only the northern hemisphere but also the southern.

Another form of reflecting telescope is the one which is the basis of today's most advanced models. This was first thought of by the Frenchman, Cassegrain, in 1672. He put forward the idea of using a convex secondary mirror instead of a concave.

In the 19th century, there were a number of notable instruments produced. James Nasmyth (q.v.), a Scottish engineer, who was already well known for his invention of a steam hammer, designed a reflector telescope in 1842: this was first used near Manchester and in 1856 was moved to Penshurst in Kent. It had a focal length of close to 14 ft and used a speculum mirror with a diameter of 20 in; the instrument was normally used with a convex secondary mirror and so was really a Cassegrain, although it could also be used as a Newtonian reflector.

In Ireland there came into being what was known as the Leviathan of Parsonstown, which when it was made in 1845 was the largest telescope in the world and held this distinction until 1919 when a 100 in specimen was built on Mount Wilson in California. It was designed and made by William Parsons, the third Earl of Rosse (q.v.), and this huge apparatus, when viewed from a distance, gave the impression of being some vast siege gun. It must have presented an odd addition to the rural scenery of central Ireland. Its mirror was 6 ft in diameter and it had a focal length of 53 ft. Its reflection system was Newtonian. The main tube was made rather like a long barrel, with wooden staves held in position with iron hoops and the whole was raised and lowered by a system of chains worked with a hand-operated windlass. Although the telescope could be raised from an almost horizontal position through the zenith to the pole, the lateral movement was small, owing to the design of the masonry walls, and it was only possible for it to be moved through one hour of right ascension. A further restriction was the weather. Nevertheless, because of the instrument's power to look further into space than any previously constructed, Lord Rosse succeeded in establishing the spiral nature of the nebulae, a discovery of fundamental importance to science.

The mounting of large telescopes calls for considerable skill and engineering knowledge. In the first place the observer must be able to manoeuvre the instrument with great delicacy and at the same time it must be completely free from untoward wobbling. To achieve these results the mechanics of the mounting had to be carried out with as much care as manufacturing the telescope itself. Modern instruments that may have pierced the obscuring distances are now more in the nature of engineering master-pieces than the earlier instruments which, besides being breakthroughs with their invention, were very often minor masterpieces of metal working and were at times considerably adorned by the skill of an engraver.

NAVIGATION

Navigation is the science or art of taking a ship across the seas or a camel across the desert. The early accounts of the subject are very vague, hidden behind legends and stories that have been passed down, embellished and have probably become highly inaccurate with the telling. Man undoubtedly sailed the seas for thousands of years before instruments to help him plot his course were available. In Europe, a great deal of the navigation would have been done by following the coastline, so that the early history of guiding a ship through the water is allied to the growth of map making and the study of geography.

It seems likely that there must have been hardy and courageous people who, planning longer voyages of exploration, turned their eyes to the heavens and picked a course with the aid of the sun, moon and stars. But no-one can tell how many of these captains wrecked themselves in fog or low visibility or became lost when the skies lay overcast for days on end.

The Babylonians were probably the first people to try to reason out the universe. They thought out the sexagesimal system, dividing the circle of the sky into 360 degrees, and the degrees into minutes and the minutes into seconds. The day was also divided into hours, minutes and seconds; this gave a relationship between the earth and the sky and allowed the earth to be plotted in relation to the stars in a co-ordinated and proportional manner. The oldest known instrument for measuring angles, the gnomon, was, according to Herodotus, invented by the Babylonians. Their concept of the world seems strange today, yet looked at with their eyes it is understandable. The area of the known earth was then comparatively small. They, therefore, looked upon the world as a flat disc that lay in the middle of some vast ocean, and outside this they believed there were seven islands which formed a bridge to an outer circle that was a mystic place for the gods.

It is possible that some wide ocean voyaging may have taken place with the aid of currents and known cycles of wind. There are theories that the Trade winds of the South Atlantic may have pushed some intrepid sailors across from the north-east coast of Africa to South America, and set currents may have operated in the same way in the Pacific. In the China Sea and the

Indian Ocean the directional steadiness of the monsoons would certainly have allowed vessels to sail on long trips out of sight of land and yet be reasonably sure they would reach close to a chosen destination. The captains could choose to go one way in one season of the year and return later when the winds had reversed.

The early Greeks had a systematic approach to map making and, although they were still working on the theory that the earth was flat, they produced examples that made some sense and were not as misleading as some of the early attempts. If the Greeks did not know about an area, they did not guess but left that part of the map blank. Homer believed that the world was shaped like a flat shield that was encompassed by a great river called Oceanus. The centre of this world he gave as Delphi. Other scholars supported these theories including Anaximander (610–547 BC) and Hecataeus (c.540–475 BC). It is recorded that Anaximander of Miletus produced a map of the world in the 6th century BC. And Hecataeus, who was also from Miletus, made a form of geographical manual in about 500 BC. In this he gave some idea of the distance between places by stating the number of days sailing between them. Herodotus wrote his history of the world in 444 BC after travelling through Babylon, Egypt and Persia.

Xenophon journeyed eastward, Pytheas of Massilia (q.v.) came to northern Europe and possibly Britain (c.325–323 BC). All these explorers were gradually pushing back the clouds of superstition and fear that obscured the horizons. Much of the information they gained was collected by Eratosthenes (q.v.), who was the librarian at Alexandria about 200 BC. In the year 20 BC, Strabo (q.v.) worked on the findings of the earlier scholars, making some corrections and adding fresh facts, and in his work mentions a terrestrial globe.

In the second century AD Claudius Ptolemy (q.v.) gave the world astronomical and geographical theories and information that were to stand undisputed for centuries. His *Geographia* was consulted as the most important volume on the subject for more than 1,000 years. Ptolemy, as a resident of Alexandria, not only had ready access to the library there but was also in a good position to consult returning sea captains, as the port was one of the most important in the Mediterranean. He also worked out the system of latitude and longitude and gave these their names. There were some faults in the master's reasoning, among which was his habit in map making of filling in areas of which he was ignorant with imaginative features. He is said to have made some terrestrial globes, but none has survived.

It is interesting to note that a number of the early astronomers were able to fix positions on the land by practical astronomy, yet it was some time before it was possible to do the same whilst at sea. The reasons for this may be twofold: instruments had to be designed that could be used on the unsteady deck of a ship; and, even if such instruments were available, sailors of those days would not have had sufficient scientific education to be able to use them.

One of the earliest leading figures in the growth of navigation was Prince Henry of Portugal, known as Henry the Navigator. Before his reign, expeditions would have sailed very much by guesswork or primitive sightings. Voyages must have been a dangerous business, especially when venturing far out from the coast. Prince Henry devoted all his energies to improving the situation. He built an Observatory at Sagres, near Cape St Vincent, in order to obtain more accurate tables of the declination of the sun. His work was continued by John II, who came to the throne in 1481. Scholars working under King John's direction improved the astrolabe so that it could be used on a ship and would be more efficient than the earlier cross-staff. The *Ordenanzas* of the Spanish Council of the Indies at this time showed that there was a prescribed course of instruction for pilots, which included Scrobosco's *De Sphaera Mundi*, Regiomontanus's spherical triangles, Ptolemy's *Almagest*, the study of cartography and the observation of the movements of the heavenly bodies, using the astrolabe and other instruments. In the 15th century, the navigational equipment for a long voyage vessel would have included a cross-staff or an astrolabe, a compass (the Chinese had a magnetic compass from about 1297), a table of the declinations of the sun, possibly a correction for the altitude of the pole star and a chart that might be more fanciful than accurate. The speed of a vessel was gauged from the 'Dutchman's Log'. This somewhat unreliable method involved throwing some object that would float out from the bows. Two observers were stationed in the waist of the ship at a measured distance apart; by noting the time the object took to pass between the two sailors it was possible to estimate the speed of the ship.

With such paraphernalia, a stout heart and surely a belief in achieving the impossible, Christopher Columbus (*c.*1436–1506) set sail for America. It is now almost certain that he must have been preceded by those intrepid voyagers the Vikings, who, braving the spray-skimmed waters of the North Atlantic, landed first in Greenland and then apparently went on, either by design or because they were swept past to the shores of Newfoundland. However, a glance at a globe, not a map, will show that the Viking voyages could have been made in comparatively short stages, and the climate was better then than now. Be this as it may, the voyage of Columbus is not to be belittled.

Columbus's early trips took him as far afield as the Levant and perhaps even Iceland. It is recorded that in 1470 he was wrecked off the coast of Portugal, near Cape St Vincent and was washed ashore on a plank from the wreck. From then on he settled himself in Portugal and within a few years he had married a daughter of Bartholomew Perestrello, a leading captain and navigator. For some years Columbus was in contact with Paola Tascanelli, a Florentine astronomer, trying to find out if it would be possible to reach Asia by sailing westward. He realized that the earth must be spherical, but he had no idea of just how large it was. With the backing of King Ferdinand and Queen Isabella, Columbus set sail from Palos on 3 August, 1492. His fleet, if such was the word, consisted of the *Santa Maria* of 100 tons and two caravels, the *Pinta* of 50 tons and the *Nina* of 40 tons. He made first for the Canary Islands and from there on 6 September headed out into the Atlantic for, as he thought, the distant coast of China or Japan. From the start he had trouble with a discontented and nervous crew who took any untoward sign as a harbinger of danger: variations of the magnetic needle of the compass were enough to render them almost useless with terror. By 12 October he had sighted an island which he named San Salvador, and which is today probably Watling Island. The expedition then found Cuba and Hispaniola (Haiti). The *Santa Maria* went aground on Hispaniola and had to be abandoned, so Columbus returned to Europe with the two caravels. In Spain he was fêted and honoured for his discovery.

He was to make three more trips across the Atlantic, with his fortunes vacillating violently. In 1499 he was actually sent home in chains when Ferdinand withdrew his favours. But after a while he was released and again honour was accorded him. The last months of his life were passed in near poverty in Valladolid.

In the latter part of the fifteenth century comes the first existing evidence of a geographical globe. This is the famous 'Erdapfel', as it was called by its maker Martin Behaim (*q.v.*). It is a remarkable construction of papier mâché covered with parchment. It has a diameter of 51 cm and is hand painted with much rich decoration. Apart from showing the countries, there are 111 miniatures, 48 flags and 15 coats-of-arms. Figures of kings, travellers and saints are included, as well as many land and marine animals. This treasure is one of the prides of the Germanisches Nationalmuseum in Nuremberg.

The science of navigation is different from any other science for one reason – the sea itself. Instruments can do much, all the knowledge of scholars can lay down the rules, but the true sailor knows only too well that his sweet savage mistress the ocean can take over in all too short a time. Barometers can foretell approaching storms but that is all. On a dark night the sea still calls the tune. No instrument yet devised can foretell the coming of those giant 'seventh' waves which can pound the largest vessel and in the case of smaller craft bring everything down to a matter of basic survival. The scientists can come up with marvels of invention, but the captain on his bridge is still the lonely one who must make the decisions when the elements start to get rough.

One of the first writers to attempt to set down rules for navigation at sea was Jon Werner of Nuremberg: in 1514 he published some notes on Ptolemy's geography, and described, amongst other instruments, the cross-staff. He said the mariner should find his longitude by measuring the distance between the moon and a star; unfortunately this advice was difficult to follow, as there was little accurate knowledge of the exact positions of the moon and stars, and no instrument that could perform this task from the deck of a ship was available. In 1530 Gemma Frisius (q.v.) wrote a treatise on astronomy and cosmogony, with the use of the globes. In this he dealt with meridians and the establishing of longitude and latitude, and with scales for measuring distances. He gave rules for finding the course and distance correctly and details for the movement of tides. This volume must have become quite well known because it was later translated into French by Claude de Bossière. In 1534 Gemma made an 'astronomical ring'. Admitting that it was not entirely his own invention, he claimed that it would do all that astrolabes, cylinders and quadrants had done.

In about 1530 watches were invented and Gemma believed that they could be used to determine the difference of longitude between two places by a comparison of local times at the same instant. He found, however, that early timepieces were too inaccurate and proposed that they should be corrected by water clocks and sand clocks. For rough timekeeping, sand glasses were used in ships and it is interesting to note that the British Navy used hour and a half glasses until 1839.

In the 16th century there was a considerable impetus to produce a workable system for the production of charts that would at least give the seaman some reasonably accurate guidelines. In 1537 Pedro Nuñez, cosmographer to the King of Portugal, published a work on astronomy, charts and some points of navigation. He saw the errors in plane charts and set out to correct them. Eight years later Pedro de Medina brought out *Arte de Navigar* at Valladolid, dedicated to Don Philippo, Prince of Spain. This is likely to have been the first book ever published to deal only with navigation. In 1556 Martin Cortes brought out, in Seville, a book with a similar title. It contained drawings of a cross-staff and an astrolabe, declination tables for the sun for four years and a calendar of saints' days. Cortes described an instrument which could tell the time, where the sun would set, and the direction of the ship's head. He also described a system of circular cards that would tell the times of high tides. Other treatises on charts were published by William Cunningham in England in 1559 and by Michael Coignet in Antwerp in 1581; also, again in Spain, by Roderico Zamorano in 1585. By far the most important of these writers and geographers was Gerardus Mercator (*q.v.*) who was born at Rupelmonde in Flanders. Early in his life he met Gemma Frisius, who inspired him to turn towards scientific geography and cartography. In 1534 he founded his establishment in Louvain. He received a commission from Charles V to make a set of instruments of observation for the monarch's campaigns, but these sadly were lost in a fire in 1546, and he was asked to prepare another set. Mercator's greatest contribution, however, particularly to navigation, was his method of projection. This is a way of map making in which the meridians are drawn parallel to each other, and the parallels of latitude are straight lines whose distances from each other increase with their distance from the Equator, so that at all places the relationship that degrees of latitude and longitude have to each other on the map is the same as they would have on the sphere itself. This produces an apparent enlargement of the polar regions, but is of great importance for navigational purposes, since a rhumb line in such a chart is always a straight line. A rhumb line is a line of constant course and therefore makes a constant angle with meridians it crosses.

After Mercator, John Davis, the famous navigator, published *The Seaman's Secrets*, a book of only 80 pages which was intended entirely for the use of sailors. The equipment he listed as necessary for a voyage included a cross-staff, a sea compass, a chart, a quadrant, an astrolabe, an 'instrument magnetical' for finding the variation of the compass, a horizontal plane sphere, a globe and a paradoxical compass.

Chart making was now becoming a science rather than an exercise in imagination. One of those who did much to make the seaman's passage safer was Edward Wright of Cambridge. He not only improved the accuracy of charts but also developed two instruments to assist in navigation. These were a quadrant that was used by two people and sea rings for taking observations. Wright was one of the first to understand how logarithms could be used to assist with navigation. Here also the mathematician Gunter (*q.v.*) was working along similar lines, and he brought out tables of logarithmic sines and tangents in 1624 and 1636, which could be applied to astronomy and navigation.

In the first half of the 18th century two instruments which together were to revolutionize navigation were invented and developed. These were the sextant and the chronometer. The credit for the sextant goes to John Hadley (*q.v.*), the English mathematician and mechanician. Although this claim at the time was challenged by Thomas Godfrey (1704–49), a glazier from

Philadelphia, investigation clearly showed that both men, whilst working along similar lines, had been proceeding quite independently of each other. Hadley's first instrument was made in 1730 and later improved. This device measured the angle between two objects. Earlier instruments for use at sea either meant that the observer had to look in two directions at once, or had to use a plumb line, which obviously was totally unsuited for operation from a wildly tossing ship's deck. Hadley's sextant could be held in the hand and the reading was taken at the moment of coincidence when one object was reflected upon another by the use of a small mirror. The instrument could then be clamped so the angle could be read off with ease and accuracy. Four years after Hadley's sextant came a chronometer, which allowed for accurate timekeeping, and thus put nautical navigation on the sound basis from which it has progressed to today's almost foolproof methods.

In 1714 the British government set up a Commission to give awards and encourage inventions and theories that would materially assist the conquest of the sea. The reason for the Commission was probably the horrific reports of nautical disasters of the late 17th and early 18th centuries. A document put together by Captain Lanoue in 1736 highlighted some of the worst incidents that were caused largely by ignorance of navigation. In 1691 several men-of-war were wrecked off Plymouth because they mistook the Deadman for Berry Head. Three years later Admiral Wheeler's squadron piled up on Gibraltar when they thought that they were safely through the Straits. In 1707 much of Sir Cloudesley Shovel's fleet smashed into the sharp rocks of the Scilly Isles and was sunk.

Up to 1828, when the Commission was disbanded, it had paid out some £101,000 in awards to inventors and those who had produced ideas that could in some way contribute to nautical safety. These included varying sums for methods of determining longitude, publishing nautical tables, and attractive offers to those who could for instance discover the North West Passage or sail close to the North Pole. The Commission further undertook to partly underwrite a thorough survey of the coasts of the British Isles. A gentleman by the name of Whiston was appointed surveyor and the records state that in 1741 he was awarded a grant for the purchase of instruments. Further afield the Commission gave sums to support Captain Cook on his scientific voyages and also assisted with the publishing of the results on his return.

Many were the almanacs, tide tables, and other publications which came into print to ease the lot of the mariner. One of the most popular of these was brought out by Nevil Maskelyne. This was the *British Mariner's Guide*, which came out in 1763 after Maskelyne's return from a trip to St Helena to observe the transit of Venus. With this book he tried, amongst other points, to give a simple method of finding the vital longitude whilst at sea by using Hadley's sextant. In 1765 Maskelyne was appointed Astronomer Royal and he put into being the system for the publication of the *Nautical Almanac*. This book soon came to be regarded as authoritative and reliable – so much so that other countries paid it the compliment of either translating it or bringing out a volume very similar in content.

The Admiralty Hydrographic Office was opened in 1795, and in that same year an order was given that a complete catalogue of all reliable charts should be prepared. An earlier one issued in 1786 for the East India Company gave details of 347 charts which covered the area extending from the British Isles, via the Cape of Good Hope, to India and China. The first official Admiralty catalogue of charts was brought out in 1830 and contained 962.

Accidents at sea still did occur and it is likely that they still will, but at least from this time the captains had some reliable systems to use and instruments which enabled them to pinpoint with accuracy their positions when pushing their way through the rolling wastes of the great oceans. Science had largely removed the guesswork from navigation.

MEASUREMENT OF TIME

The basic measurement of time for early man can have been little more than an unconscious or conscious noting of the passing of each day by the successive appearing of light and darkness. One of the first attempts to get some kind of regularized time measurement was by the Babylonians. They had a calendar of twelve lunar months, but this had disadvantages as it was based on the fact that in the year there were $12\frac{1}{3}$ lunar months, each lunar month being $29\frac{1}{2}$ days long. They got rid of the half day by having alternate months of 29 and 30 days long. The lunar year, however, is 11 days shorter than the solar or natural year. The Babylonians tried to straighten this out by adding an extra month every third year but, for all their efforts, over a long period of time small errors built up and confusion resulted.

The early Egyptians produced a calendar which was based on the sun. This had 365 days, which were split up into 12 months of 30 days each with five days added at the end of each year. The Greeks divided their month of 30 days into three equal divisions, a method that revolutionary France later tried to follow.

The early Roman calendar seems to have been one of the most confused of all. It had ten months, the days of which were calculated backwards from the three fixed periods of calends, nones and ides. This calendar has been ascribed to Romulus and had the defect at first that it left out 60 days in winter altogether; the year started with March and ended with December but the two missing months of January and February were added in the reign of Numa. All the months were then arranged to have 29 and 30 days alternately, but this still gave a short year of only 354 days. One extra day was added, for the sake of superstition as odd numbers were deemed lucky, to make 355 days. Still the calendar was ten days short and to get round this an extra month was inserted every two years. This extra month was alternately 22 and 23 days, which made the year one day too long. The final adjustment was in no way scientific, it was left to the rulers to wield as an instrument of power, curtailing the year to spite their enemies and lengthening it to benefit themselves and their friends.

By the time Julius Caesar became dictator he found that the calendar was in chaos. One of the first tasks he set himself, with the help of an astronomer, Sosigenes from Alexandria, was to set up a new calendar. Caesar ordained that the average length of the year should be $365\frac{1}{4}$ days. Every fourth year would have 366 days and ordinary years would have 365 days.

The first year of this, the Julian Calendar, was 46 BC. Its most important advance was that this calendar accounted for the odd quarter day difference between the calendar and solar year. As a mark of recognition for Caesar, just after his death the name of the seventh month of the year was changed from Quintilis to Julius (July).

The first Emperor of Rome, Augustus, made some alterations in the year 8 BC. He took a day from February and added it to August, a day from September and added it to October, and a day from November was given to December, which brought the calendar to the form that has been followed since. Sextilis, the month following July, was renamed Augustus (August) to honour the Emperor. The reformed Julian calendar with the four-yearly leap year was not accurate enough, however, as the year was still too long by 11 minutes and 14 seconds. In the early years of the system this quite small amount would hardly have been noticed, but by 1580 the first day of spring was some ten days early.

In an arbitrary attempt to rectify matters Pope Gregory XIII in 1582 ordered that ten days should be dropped from October to readjust the calendar and 4 October was followed by 15 October that year. To make a permanent adjustment so that this error should not arise again it was ordained that the centurial should not be recognized as leap years, except where they were divisible by 400. Thus 1600 was a leap year, but 1700, 1800 and 1900 were common years and 2000 will be a leap year. This method of calculating the year was called the Gregorian Calendar and is the one that is still used today. The calendar was promulgated by a Bull and it was accepted in France,

Spain, Portugal and parts of Italy. Scotland did not adopt it until 1600, and most of the German states took it up at the end of the 17th century. England held out against using it until 1752. By this time the difference between the Julian and Gregorian calendars had grown to 11 days, so in England for that year 2 September was followed by 14 September. At the time this act caused an outcry: working men thought that they had been cheated of 11 working days and tried to insist that they were paid for them.

One of the earliest attempts to divide periods of daylight was made by the Egyptians with their shadow clock. They had noted how shadows move across the ground in relation to the movement of the sun across the heavens. The Egyptians' shadow clock consisted of a 'T' made from metal or wood, the top part of the 'T' being bent up at a right angle so that if it were aligned facing the east in the morning a shadow would creep along the horizontal part of the 'T' and a rough timing could be given as it crossed a system of marking. In the afternoon the shadow clock had to be realigned facing the west; there were graduations to show six hours that could divide up a 12-hour day for the Egyptians. The hours would of course have to be further regulated between summer and winter to allow for the longer and shorter times.

The development from the shadow clock was the sundial. There have been many variations of this but they are all basically the same in principle. The standard fixed face version has a pointer, which is called a gnomon, in the centre, so angled that when it is correctly aligned to the axis of the sun the shadow will move across a scale and so tell the time. The Romans used sundials and so did the Saxons, although their models were marked off into four equal periods called tides. The terms 'noontide' and 'eveningtide', still used today, are derived from this Saxon method of recording the time.

The Babylonians and early Egyptians, and the Chinese, had water clocks, which were also used by the Greeks and Romans. Basically these devices were simple. They worked on the principle of a container with a small hole in it which allowed the water to leak out; as the level fell it could be measured against some form of scale. One of the earliest types was a copper vessel, which had a small hole in the bottom; this was floated in a tank of water. When it sank it signified the passage of a certain period of time. The Greeks and Romans are known to have used water clocks to limit speeches in their courts and it is from the Greek that the classical name for a water clock, clepsydra, comes. Some quite sophisticated models were connected to rods or pulleys which rotated a pointer that moved over a graduated dial.

Another somewhat unreliable device for telling the time was the candle clock, in which candles were marked off with graduations to show the passing time, which was indicated as the flame burnt lower and lower. The accuracy of candle clocks depended very much on how carefully the candles were made. Even a very small difference in the diameter could markedly affect the time telling, and the composition of the wax could affect the rate of burning. The Chinese had a refined version of the candle clock in which the flame travelled along a horizontal wick and at set intervals burnt through thin threads on which were hung small metal balls, which fell and produced a strike for the period of time.

For short periods of timekeeping sand glasses have been used for a long time. The principle again is very simple; the sand glass consists of two glass bulbs joined together with a narrow constriction or with their necks in contact with a small metal plate in which a tiny hole has been drilled. An amount of dry sand is put into the glasses during manufacture and this gradually trickles from one bulb to the other. Sand glasses are sometimes mounted together in sets of four, which will give the passage of quarter, half, three-quarters, and the hour. They were useful in ships because the movement by rough seas would not materially affect their operation. As with water clocks, they were also used to limit speech makers, and preachers in church.

The earliest 'proper' clocks appear to have been in use in Europe during the 13th century. There is, however, evidence that they may have been invented several centuries earlier, though these examples are likely to have only been curiosities. The first clocks were balance clocks and their invention has been attributed, in a roundabout way, to Pope Silvester II in about 996. These old clocks had a verge escapement and a balance. The train of wheels ended with a crown wheel, that is, one with serrated teeth like a saw, placed parallel with its axis. The teeth engaged with pallets mounted on a verge or staff, placed parallel to the face of the crown wheel. As the wheel turned, the teeth pushed the pallets alternately until one or the other slid past a tooth and thus let the crown wheel rotate. When one pallet had slipped over a tooth the other pallet caught a tooth on the opposite side of the wheel. The verge was connected to a balance rod, placed at right angles, and had a ball at each end. The accuracy of such clocks was very much affected by friction and lubrication of the moving parts was essential.

In 1582, Galileo first discovered the properties of a swinging pendulum. He noted that the swing did not depend on the weight at the end of the pendulum but on the pendulum's length. It is not known whether he or his son Vincenzo ever actually made a pendulum clock, although there is a drawing of a mechanism for such a timekeeper. It was apparently Hevelius (q.v.) of Danzig who first took advantage of the breakthrough and made two such clocks after reading about Galileo's discovery. In 1656 the Dutch scientist Huygens (q.v.) demonstrated that the earlier non-pendulum clocks could be quite simply adapted to pendulum control.

Watches were first made in Germany and France during the first half of the 16th century but no English craftsman appears to have been operating before the beginning of the 17th century. One of the first of these was Richard Jackson of London. The development of the chronometer, which is really a very accurate watch, was of the greatest assistance to the navigator. A pioneer in this field was John Harrison (*q.v.*), the son of a Yorkshire carpenter, who spent much of his early life in Lincolnshire, and became a leading horologist. In 1714 the British government offered a £20,000 reward for anyone who could produce a timekeeping device that would only show an error of two minutes or less on a voyage to the West Indies. Harrison actually made four models. The first three were quite large and cumbersome, but his fourth was in the form of a watch about 5 in in diameter. This was put on board HMS *Deptford* in 1761 and on arrival in the West Indies showed an error of only five seconds. Despite this triumph, Harrison had to wait until 1773 before he was paid his reward.

A famous instrument that should be mentioned with regard to the regulation of time is Airy's (*q.v.*) transit circle, on which more than $\frac{3}{4}$ million observations were made between 1851 and 1954, the Greenwich Meridian was defined by the International Meridian Conference of 1884 as 'the Meridian passing through the centre of the transit instrument at the Observatory at Greenwich'. When the Astronomer Royal and his staff moved after the Second World War to Herstmonceux, Sussex, Airy's transit circle was left at Greenwich, and although Britain's time signals now emanate from Herstmonceux, the Greenwich Meridian remains the basis of the International Time Zone System.

LINEAR MEASUREMENT

It is not known exactly when man first began to use a linear measuring system. It may have been as early as 8000 BC, but it is likely to have been simply a primitive pacing out of the dimensions of land. The roots of ancient metrology have many gaps still and the task of unravelling the various branches of the subject is made difficult by legend and theory.

The first degrees of measurement were derived from parts of the body. The smallest indication used was a digit, which was the breadth of one finger. Four digits made a palm or handbreadth. A span was 12 digits or three palms; the foot was equal to four palms or 16 digits. The Jews used either the length or breadth of the top joint of the thumb as a measurement and specified which. The measurement most often used was the cubit, which was 24 digits or six palms. The height of a man is commonly quoted as being four cubits. By this recording Goliath must have been truly a frightening figure as he is reputed to have stood six cubits and a span high; this would have made him a little over 9 ft 7 in. Another tall man in the Old Testament was Og, the Amoritish king of Bashan, whose bed, according to *Deuteronomy*, Chapter 3; verse 11, was nine cubits long and four cubits wide. This makes the iron bedstead more than 13 ft 3 in long and nearly 6 ft wide.

In the *Rhind Papyrus*, which is in the British Museum, there is a reference to a rope stretcher. This was a person who took measurements for buildings and also measured land. His chief tool was a length of rope which had knots at intervals of one cubit. This could be used in the same way as the surveyor's chain, which came later. For measuring land a rope with knots at intervals of five cubits may also have been used.

The first attempt at a master standard appears to have been made in about 3000 BC in Egypt. Here the old cubit of 18 in had been slightly lengthened to 20·63 in and was known as the royal cubit; it appears to have been strictly adhered to in construction and building work. In hieroglyphic picture symbol writing, the symbol for a cubit was a forearm. For longer lengths, multiples of the cubit were used. The smallest of these was the *xylon* of three cubits which was about the length of two paces. Then there was the *net* or fathom of four cubits, the *khet* of 40 cubits and the long distance marker of 12,000 cubits, known as *schoenus*, which was just under 3·5 miles.

In Assyria and Babylon the standard cubit varied slightly and was between 20·5 and 20·6 in. To assist in taking smaller measurements the span, which was about 10·5 in, and the digit of 0·653 in, were used. The Sumerians employed mostly a double cubit which was about 39 in. It was divided into 60 fingers (*su-si*), with intermediary units of the half-foot of 6·5 in, the half cubit (great span) of 9·75 in, the foot, of 13 in, and the cubit of 19·5 in. For land measures they used the reed or rod of 9·74 ft, the pole (*gar*) of 19·5 ft and the cord of 195 ft. In the Louvre there is a grey basalt statue of Gudea, Prince of Lagash, *c.*2170 BC, on whose knees is a drawing tablet with a plan of a temple and a scale showing 16 finger divisions of the Sumerian cubit.

The early cubit measuring rods were of stone, rectangular in shape with the measuring graduations marked on a bevelled edge. Stone must have been too cumbersome as wood quite soon replaced it. In order that the accuracy of the wooden rods could be watched, a standard measure would often be cut into the wall of a temple.

From the Egyptian royal cubit a number of measures have been derived, including the common Greek foot of 12·45 in, the Etrurian foot of the same length, the foot used in England up to the 12th century, of 12·47 in, and the Rhineland foot of 12·36 in, which was the last unit of the Prussian system in 1816, prior to metrification in Germany.

There is yet another cubit that is particularly associated with the Teutonic races and this is known as the Northern Cubit of 26·4 in, which gives a foot of 13·2 in, although the foot when used alone was generally given a size of 13·1 in or slightly more. The use of these measures can be traced from 3000 BC until the introduction of the metric system during the last century.

The foot measure is one of the most widespread and oldest of the linear units and variants of it, or of the northern cubit from which it comes, occur in widely diverse areas. The Chinese foot has 13·2 in, a foot from Moscow is 13·17 in, and one from Verona 13·4 in; there are examples of the full cubit as used in Antwerp of 26·96 in, in Lisbon of 26·7 in, and Nuremberg of 26·0 in.

Other cubits included the royal Persian with a length of 25·2 in, which was the measure of Darius the Great (521–486 BC), and the so-called black cubit of the Arabs which was 21·28 in.

The Roman foot owes part of its derivation to the Egyptian *remen*, which was 11·65 in in length. The complete Roman measurement system was the *digitus*, $\frac{1}{16}$th of the Roman foot (0·73 in); *uncia*, $\frac{1}{12}$th of the foot (0·97 in); *pes*, the foot; *palmus* (2·91 in); *cubitum* (17·48 in); *passus* (4·86 ft); *decempeda* or *pertica* (9·71 ft), *actus* (116·54 ft); *stadium* (606·9 ft); *mille passus*, the roman mile (4,856 ft) and the *leuga* (7,283 ft). Roman rulers have been found and a good example is preserved in a London museum. This is hinged in the centre and has a clamp to hold it steady when unfolded. On one side it is marked off with 12 *unciae* to make a roman foot of 11·65 in and on the reverse it is marked with 16 *digiti*.

For the Anglo-Saxons, adequate means of measuring land were important, for they were a people very much concerned with farming. During the reign of King Athelstan, which lasted from 925 till 940, some new measurements were introduced. The full scale read like this: the Northern foot was divided into four palms of 3·3 inches, as well as twelve thumbs of 1·1 inch or thirty-six Barley corns of ·37 of an inch. They still used the cubit, but it was now called an *ulna* and measured six palms long. The thumb is an interesting measurement, as still, at times, a carpenter can be observed using the flat of his thumb to mark off inches. For longer distances, the scale included a land rod, made up of 15 northern feet and a furlong or furrow length of 40 land rods, giving it a length of 660 ft. Both of these measurements have maintained their length and still exist today.

It was William I who ordered the first full survey of England. This was begun in 1085 and was recorded in the *Domesday Book*, in which all the holdings of land are measured in perches, roods, acres and hides. The hide was 160 acres and was made up of 4 *virgata* of 40 acres each. The *virgata* consisted of 4 *ferlingata* of 10 acres each, a section of land which would be a furlong square. At this period comes the first mention of the yard. One version of its derivation is that it was the length of the arm of King Henry I, another is that it was the length between the eye and the end of the outstretched arm. A third possibility is that it was related to the length between the hand holding the drawn bow and the place where the other hand rests near the eye when the arrow is drawn back for shooting.

In the reign of Henry III, the Great Charter of 1216, a revision of Magna Carta reissued by the great barons of the realm who were acting as guardians of the infant king, amongst other points made a start at getting a standard measure of length for the country. In one place it specified that, for cloth measurement, there must be two *ulnae* between the edges of the material. This would give a width of 52·8 in. The range of the foot in Britain at the time was quite formidable, as it included the Pythic or natural foot which was used in Wales for measuring land and was of 9·9 in, the Roman foot of 11·65 in, the northern foot of 13·2 in and the Greek common foot of 12·47 in. In 1305 a statute issued by Edward I read:

> It is ordained that three grains of barley dry and round make an inch, twelve inches make a foot, three feet make an ulna, five and a half ulna make a rod, and forty rods in length and four in breadth make an acre.

Between the beginning of the 14th century and the end of the 16th a number of standard yards were produced. One of the earliest of these may have been first stamped in the reign of Edward I: made of bronze, it was stamped again in the reign of Henry VII and is now in the Westgate Museum in Winchester. Compared with the Imperial standard yard of today, it is 0·04 in short. In about 1497, during the reign of Henry VII, a bronze standard, encased in silver, was made for the Merchant Taylors' Company; this was 0·006 too long by the present standard. In 1588 came the standard yard of Queen Elizabeth I, which lasted until 1824; this was 0·01 in short and is today kept in the Science Museum. The Elizabeth standard is a bronze bar of 0·6 in square section with the measurement of the yard being taken from end to end of the bar, whilst engraved lines mark off a sixteenth, an eighth, a quarter and a half. In Elizabeth I's reign the English mile is first mentioned. This was of eight furlongs, 5,280 ft or 1,760 yards, which is the same as it is today.

WEIGHT

As with linear measurement, it is difficult to pinpoint the exact moment in history when man started consciously to try to weigh something against a prescribed set of standards. Some limestone weights dating back to about 3800 BC were discovered in graves at Naqada in Egypt. They would, presumably, have been used with some form of simple balance. The main unit of weight in early times was the *shekel*, which varied in the same way as the cubit. *Shekels* were made up from grains, which averaged about 200 to the unit. The principal standard of weights was the *beqa*. This lasted for 3,000 years in Egypt and was then taken up by the Greeks. From it developed some of the Roman pounds; it was also used by the Arabs and was the foundation of the English Troy weight. The *shekel* mentioned in the Bible is likely to have been one that weighed 224·5 grains; smaller than this was the *bekah*, the half-shekel, and the *rebah*, the quarter. Further up the scale were the *libra*, the pound that is mentioned in *John*, Chapter 12 verse 3, the *mina* of about 2 pounds Troy weight and then the *talents*. The *talents* were either silver, of 117 Troy pounds, or gold, of 131 Troy pounds. The origin of the term grain is that it was supposed to represent the weight of one grain of wheat.

The idea for the first rather primitive scales probably came from the poles by which the Egyptians used to carry loads. These were balanced across their shoulders, as can be seen from drawings in the tombs. There is a low relief panel in a Fifth Dynasty tomb at Gizeh that clearly shows two figures with a cross-bar scales, in which the whole is supported on a stand on the floor.

An early civilization with an advanced system for weighing occupied the two cities of Harappa and Mohenjo-Daro in the Indus valley about 2000 BC. Sir John Marshall, director of archaeology in India in the 1920s, carried out excavations which showed that the people of these cities had lived a remarkably advanced life. Their cities were laid out on a grid-iron plan of streets with the buildings themselves showing considerable architectural skill and such refinements as bathrooms and drainage systems. At Mohenjo-Daro, there is a bath, some 39 ft long and 23 ft broad, built of brick with gypsum mortar and a thick bitumen damp-proof course. Weights found here were made of a hard flint-like material called chert. Although the hundreds of weights ranged from about 13 grains up to over 24 pounds avoirdupois, none of them bears any inscription or numerals to aid identification. An interesting point is that these people must have been very conscious of the importance of weighing whatever they were trading in or buying, as weights were found there in nearly every type of building.

The Sumerians of about the same period as these Indus people had a somewhat different culture. The God of the temple was regarded as the all powerful and a tythe of all the produce was brought by the farmers and others to the temple. Records exist that show in detail what goods were brought for offerings and how they were accounted and weighed. The Sumerian weight scale was again centred on the *shekel*, which in this case was of 129 grains, with a smaller weight of three grains which was called a *seu* and larger weights of the *mina* and the *talent*.

The Sumerian *shekel* was probably based on the weight of 180 grains of wheat taken from the middle of the ear. This idea of taking a grain of corn as the starting point for a weight scale is very old and occurs in many different parts of the world. Edible grains of many varieties have been chosen, including not only wheat but also barley, millet and rice. The grain sizes vary between types, and also they depend for weight on which part of the ear they are taken from; weather conditions at the time of the harvest can further affect the size.

Standard Sumerian weights have been found carefully inscribed with weight and origin; the earlier examples were generally in the image of a duck and carved from basalt or similar material. Dating from about 2400 BC, one is marked: '15 Shekels of the God Nin-gir-su, made by Uru-ka-gi-na, king of Gir-su', and another from 2095–2048 BC, the reign of Dungi of Ur, bears the legend: 'To Nannar his Lord has Dungi the mighty man, King of Ur, King of the Four Lands, two Minas of correct weight dedicated'. Later Sumerian weights lost their zoomorphic image and were plain, barrel shaped and made from haematite.

The Babylonians, those romantic and gigantic builders, carried out many schemes for defensive works, construction, temples and civic buildings. Their great city, Babylon, lay astride the river Euphrates. Here were the tower of Babel and the famous 'Hanging Gardens'. They continued with the Sumerian standard weight and reverted to the duck form, weights being carved from not only haematite but also alabaster, onyx and quartz.

One who did as much as anyone to elucidate the early Egyptian system of weights was Sir William Matthew Flinders Petrie (1853–1942). Petrie was to some degree the pioneer of new and enlightened methods for excavation, which should be carried out to a plan and in an ordered fashion, as opposed to the hit and miss methods many had used previously, where sites had been attacked with shovels and picks, and little or no note had been taken of where or how specimens were lying, and the earth had generally been removed without being even sifted properly so that many a precious fragment could be lost. Petrie built up a personal collection of some 4,000 stone weights. The earliest examples were limestone cylinders with rounded ends. After this the shape became conical and then rectangular slabs of basalt. Later still cushion shapes appeared. The weight standards of the early Egyptians included the *beqa* and also the *quedet*. The former was the most important and the *beqa shekel* was 200 grains, although there was a variation of about 12 grains either side of this.

The Greeks had two standards: the *aegina*, based on the Egyptian *beqa*, and the *attic*, based on a unit of one hundred and thirty-five grains. Confusion often arises with many of these early weights, as there is a close relationship between them and coinage. A clear example is the *drachma*, a small Greek coin equivalent to about $\frac{1}{600}$th part of a *talent*, or $\frac{1}{100}$th of a *mina*; the *drachma* itself was made up of six *obols*. As a weight measure a *drachma* was equal again to about $\frac{1}{100}$th part of a *mina* (which itself was about a pound).

The first Roman weights were produced some time after the Greek weights, one reason for this being the lack of native precious metals. Their earliest forms of money were cast from bronze and bore the heads of deities and the value on one side and the prow of a ship on the other. The weight system that they developed and used was two *obols* equalled one *scrupulum*; 24 *scrupula* equalled one *uncia* (ounce); 12 *unciae* equalled one *libra* (pound). The coin weight scale used for gold and silver was 24 *siliquae* equalled one *sextula*; six *sextulae* equalled one *uncia* and 12 *unciae* equalled one *libra*.

The Romans began weighing with the equal arm balance of the Egyptians. They then invented the bismar or steelyard. Aristotle mentions an appliance of this type in his *Mechanica*. There is, however, no accurate recorded information that this was ever used by the Greeks. The common or Roman steelyard was basically a lever where equilibrium was obtained by varying the distance of a weight from the fulcrum instead of changing the weight. One variation allowed a hanging hook to be slid along the lever, instead of moving the weight. In both cases graduations on the long arm showed the weight of the body.

The steelyard in one form or another occurred over a wide area. Versions were used as far afield as Malaya, China and India. In Europe its use was almost universal and in Russia and Scandinavia refined types were in use until quite recently. The compass of the steelyard principle was wide: it stretched from the small examples intended for weighing gold and silver up to some huge specimens, at the turn of the 17th century, which could weigh loaded carts of over 5 tons. The basic idea of the steelyard still operates in modern baggage scales and weighbridges for lorries and railway trucks.

In the 5th century, the arts and sciences of Europe were subjected to a severe setback as the advances of the Roman Empire were torn apart by King

Alaric and his Visigoths, to be followed later by King Genseric and his Vandals. The depredations of these two in particular reduced the order of the Romans to chaos and put Europe into the Dark Ages that were to last for about 500 years.

After the Hegira in 622 the Arab peoples were in the ascent and were preserving and advancing their education and arts. At the courts of leaders like Harun al-Rashid (786–809) and Caliph al-Mamun (813–33) in Baghdad, scientists and scholars could flourish. Their weight scales were based on gold and silver coinage and can be traced from the old *shekel*. They also used the *dinar* for gold coinage and the *dirhem* for silver. The word *dirhem* was used for a weight as well as for a small silver coin.

In Britain the Anglo-Saxons began to restore a form of order and progress. It was particularly under King Offa that a number of advances were made. He brought in the silver penny of 22·5 grains, based on the Arab half *dirhem*. For weighing there came in an ounce, of 20 pennyweights, and a pound of 12 ounces.

It was Edward III who in 1340, brought in the English Avoirdupois weight system, which was intended from the start to be a scale for weighing goods and would have no connection with coinage. He included in his reforms a plan for standards for not only weights but also volume measures and ordered that these should be put into universal use throughout his kingdom.

Under Queen Elizabeth I, as with linear measures, weights were further organized so that the standards that appeared were to remain until the Weights and Measures Act of 1824.

VOLUME

With the early peoples there is still some uncertainty as to exactly what measurement they may have used for bulk liquids or materials. One of the first recognizable systems appeared in Egypt with a measure called a *hon*, which held 0·84 of a pint. There was also the Syrian *kotyle*, used in Egypt, which held 0·62 of a pint. The Babylonians had the *log* or 0·95 of a pint whilst larger measures included the Phoenician *kor*, which held just over 10 bushels. The Greeks used a *kotyle*, which was about half the Babylonian *log*. The Romans had their *sextarius* of just over a pint and the *congius* of just over 6 pints. In general the systems for volume control were not as stringent as for weight and linear measurements.

In England Henry VII in a statute of 1497 introduced some order to volume measurement. His Winchester standard ordained that one pint was to contain 12·5 Troy ounces of wheat; one quart (2 pints), 25 Troy ounces of wheat; one pottle (4 pints), 50 Troy ounces of wheat; one gallon (8 pints), 100 Troy ounces of wheat; and one bushel (64 pints), 800 Troy ounces of wheat. These units were to be used for measuring ale, wine and wheat.

The gallon, after this, was altered several times. The wine gallon of the 18th century was derived from an earlier edict by Henry III which ordered that the gallon measure was to contain 8 pounds weight of wine, the pounds being made up of 'Tower' ounces. The Tower weight gave place to the Troy, which meant that 15 Troy ounces would be slightly more than the earlier measure. An Act of 1707 clarified the situation and also set out the contents for much larger measures. The hogshead should contain 63 gallons, the pipe 126 gallons and the tun 252 gallons.

The Weights and Measures Act of 1824 finally laid down a very exacting measure for the gallon. This was that it should contain 10 avoirdupois pounds of distilled water under specified barometric and temperature conditions. The Act also rendered obsolete earlier volume measures such as the ale and wine gallons and the Winchester unit.

TEMPERATURE

Serious attempts to take measurements of temperature begin with Galileo and his thermoscope. This consisted of a glass bulb containing air which was connected to a small bore glass tube that was dipped into a container of coloured liquid. This device registered changes of temperature but was unsatisfactory because it was also affected by atmospheric pressure. Later Galileo is credited with making an instrument that is closely allied to the present-day thermometer. In 1612 a fine glass tube was hermetically sealed to a bulb containing alcohol, and gradations were marked on the tube by fusing small spots of enamel. The difficulty arose in finding fixed points from which a workable scale could be devised.

In 1701 Sir Isaac Newton put forward the idea that a scale should be made on which the freezing point of water should represent zero and the temperature of the human body should be 12 degrees. In 1714 Gabriel Daniel Fahrenheit, the German physicist, suggested that zero should be taken as the lowest temperature obtainable with a freezing mixture of ice and salt and the distance between this and the human body be divided into 12 degrees. To achieve greater accuracy this number was later increased to 96 degrees.

Fahrenheit substituted mercury for the alcohol in the tube and also finally fixed a scale which had the freezing point of water at 32 degrees and the boiling point of water at 212. Besides this the Centigrade scale, invented by Celsius, with freezing point as zero and boiling point at 100 degrees, was used. A third scale known as the Réaumur, which again has freezing as zero but puts the boiling point at 80 degrees, was introduced by René Réaumur (*q.v.*), the French scientist known as the 18th-century Pliny for his diversity of study and achievement, which ranged from insects to metals and from rivers to precious stones.

Today Centigrade has been adopted as a practical international scale for temperature; a parallel development is the approaching universal use of the metric scale for linear and area measurement.

The repeated concentration of searching and trained scientific minds over the centuries has gradually but firmly pared away the unnecessary, the doubtful, and left the theories and practice clear and workable.

ENTRIES

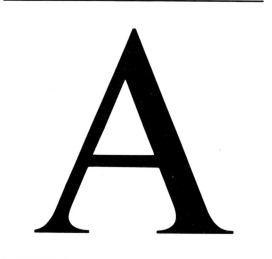

Abacus

A counting device, also used for working out mathematical calculations. Basically it consists of a series of rods in a frame, each rod designating a denomination – units, tens, hundreds, thousands in the decimal system. On the rods are specific numbers of discs representing digits; the discs might be made from bone, small stones, wood or coins. The design and capabilities have varied between countries. The Roman abacus generally had seven long and seven short bars or rods, the long rods carrying four discs, the short just one. The *Swan-Pan* abacus of the Chinese resembles the Roman in construction, wires often taking the place of rods and small balls of bone or ivory being threaded on to them in a similar manner to the bead abacus used during this century for teaching arithmetic in junior schools.

The name abacus is also applied to an instrument often called the 'logical machine'. This is constructed in such a way as to show all the possible combinations of a set of logical terms with their negatives and also to show how these combinations are affected by the addition of attributes or other limiting words in order to simplify mechanically the solution of logical problems.

Aberration

The apparent variation or displacement of a star or other heavenly body due to the motion of the observer's eye with the earth. In optics the word has two special applications: Aberration of light and Aberration in optical systems. Aberration of light can be defined as an apparent motion of the heavenly bodies – the stars describe annually orbits more or less elliptical in shape – which varies according to the latitude of the star: consequently at any given moment the star can appear to be displaced from its true position. This seeming motion is caused by the finite velocity of light. The discovery of the aberration of light was made in 1725 by James Bradley, who was born at Sherborne in 1693. A friend of Newton (*q.v.*) and Halley (*q.v.*), whom he succeeded as Astronomer Royal in 1742, he died in 1762.

Aberration in optical systems, as with lenses or mirrors or series of them, can be defined as the non-concurrence of rays from the points of an object after transmission through the system. This can happen when the image formed by such a system is irregular and it is of the utmost importance that this aberration is recognized and corrected by the instrument maker.

Accumulator

A rechargeable device for storing electrical energy in the form of chemical energy in one or more secondary cells. When the cells are discharged the chemicals react with one another and release the electric force. A hydraulic accumulator was developed by Lord Armstrong (1810–1900). Other types have included the Planté cell, the Faure, the Tudor, the Hart and the Hatch. Edison (*q.v.*) worked on an alkaline accumulator which was primarily intended for traction work.

Achromatic Lens

Achromatism in optics signifies the property of transmitting white light, without decomposing it into the colours of the spectrum. Lenses which have this ability are termed achromatic, which implies that the chromatic aberration has been corrected; this is done by assembling a pair of lenses, one of flint glass and the other of crown glass. The flint will correct the dispersion of the crown.

Achromatic Telescope

One that is made with series of lenses corrected for aberration. This is the kind normally used either for astronomical or terrestrial observations. The expense of making large instruments of this type can be considerable. No less than four surfaces have to be ground and polished to the required curvature for the object lens alone. There have been instruments made with the object lenses composed of thin glasses and then filled with transparent liquids, with the aim of saving considerably on costs, but leakages and currents set up within them by changes of temperature have prevented the idea coming into more general use.

Acorn Clock

A type that became popular in New England from around 1825. This American shelf clock usually stood about 2 ft high and had the top part shaped something like a large acorn.

Act of Parliament Clock

Late 18th century cheap English wall clock, generally weight driven with a seconds pendulum and often an unglazed dial. Supposedly these became popular when Pitt introduced an act in 1779 which put an annual levy of five shillings on all clocks and watches. The Act of Parliament clocks used to be hung in inns to assist those people who had sold their timepieces to avoid paying the levy. Pitt's act was so unpopular that it was repealed in 1798.

Adams, George and Son (active 18th century)

Father and son, both called George, were superb 18th-century craftsmen. Between them they produced a large number of different instruments including barometers, compasses, micrometers, microscopes, pantographs, prisms, telescopes, theodolites and zogroscopes. In 1746 the father made his 'New Universal Single Microscope', which incorporated the then advanced idea of a system of single lenses of varying power that were set in a disc that could be revolved under the barrel. George Adams senior also made what is still probably one of *the* fantasy instruments – an elaborate silver microscope lavish with baroque decoration.

It was presented to the Prince Regent, who presumably used it as an entertainment, in the fashion of the day. A similar instrument was made for George III. The former is now in the Science Museum, London, the latter is in Oxford. Both father and son held the appointment of mathematical instrument maker to the King.

Rare 18th century brass surveyor's equinoctial inclining dial signed G. Adams, London

Adams, John Couch (1819–92)

British astronomer born at Laneast, Cornwall and educated at St John's College, Cambridge. While he was still studying there he became fascinated by unexplained irregularities in the motion of the planet Uranus and felt that these could be caused by the action of some remote undiscovered planet. In September 1845 he had worked out a form of solution which he handed to Professor Challis, the director of the Cambridge Observatory; the paper gave the elements of what Adams described as a new planet. In October he sent a copy to the Greenwich Observatory for the attention of the Astronomer Royal, Sir George Airy (*q.v.*), who queried the young man on some of the points raised, but as Adams did not answer, no telescope search for the unknown planet was instigated. Meanwhile the French astronomer Leverrier presented a memoir on the theory to the French Academy. The discovery of the unknown planet, to be named Neptune, was made in the following year from the Berlin Observatory. Other advanced work concluded by Adams included research into gravitational astronomy and terrestrial magnetism. He worked for many years arranging and cataloguing the huge collection of Newton's (*q.v.*) unpublished mathematical writings.

Aerial Telescope

An instrument without a closed-in tube. Attempts were made to construct large telescopes of the Galilean type but these met with difficulty. The magnified image lacked purity and the edges exhibited rainbow colours. To endeavour to correct this, telescopes of considerable length but only a few inches wide were produced. These instruments were extremely unwieldy and some makers resorted to just a skeleton tube or even to no tube at all, by fixing the object glass to the top of a high post and keeping the eyepiece in line with a cord manoeuvred by the observer on the ground. From the magnification point of view the presence or absence of a tube makes no difference, the dark channel of the tube only helps to keep out stray lights and to clear and define the image. The Romans knew nothing of telescopes but tradition has it that Julius Caesar surveyed the coast of England before crossing by looking through a tube with a blackened interior which enabled him to have a clearer view. Some of the giant aerial telescopes that were made include one 123 ft long, presented to the Royal Society by the Dutch astronomer Huygens (*q.v.*); Hevelius (*q.v.*) of Danzig made one 150 ft long and Bradley one of 212 ft in 1722 to study Venus.

Air Pump

A device for removing, compressing or forcing a flow of air. One of the first

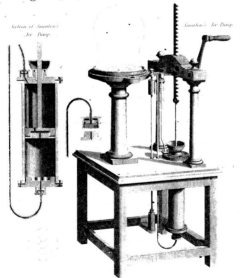

An air pump invented by John Smeaton, mid-18th century

examples was devised by Robert Boyle (*q.v.*) and described in his *Philosophical Works*; by it Boyle was the first to prove that air was essential to animal life.

Airy, Sir George Biddell (1801–92)

Born at Alnwick, he was educated at Cambridge where he became senior wrangler in 1823. After becoming a fellow of Trinity, he was appointed Plumian professor of astronomy in 1825 and in 1835 he became Astronomer Royal, holding the post until 1881. He did much to modernize the equipment at the Greenwich Observatory and started the regular recording of sunspots. One of his most remarkable achievements was the determination of the mean density of the earth. In 1826 he first had the idea of solving this problem by using a pendulum in a deep mine. His first attempt was at Dolcoath in Cornwall but this failed because of trouble with the pendulum and also the mine flooded. Later, however, at Harton Pit near South Shields he established that the gravity at the bottom exceeded that at the top and from these investigations he was led to a final value of 6.566 for the mean density of the earth.

Alfonsine Tables *see* Tables

Alidade

A rule which is equipped with sights and used in surveying for the determination of direction. Also, that part of any optical, surveying or measuring instrument which comprises the indicator, verniers, microscopes etc or the movable index of a graduated arc, used in the measurement of angles. The index-arm of an astrolabe, quadrant or other graduated instrument. The word is also used to designate the supporting frame or arms carrying the microscopes or verniers of a graduated scale.

Allen, Elias (c.1606–54)

Skilled instrument maker who produced not only popular examples but also collaborated with scholars, notably with the mathematician William Oughtred (q.v.), in their experiments to produce new types of instrument.

Almacantar (also Almucantar)

The name derives from the Arabic for a sundial. The term is given to an instrument invented in 1880 by S. C. Chandler to determine latitude or correct a timepiece. It was of great value because it was remarkably free from instrumental errors. Also, an astronomical term for a small circle of the sphere parallel to the horizon. When two stars are in the same almacantar they have the same altitude.

Almagest

A work on astronomy compiled in the 2nd century AD by Ptolemy (q.v.). The work dealt in detail with the distance and motion of the sun and moon, their planetary conjunctions and oppositions and eclipses. It contained a description of the geocentric system of the universe and a catalogue of stars.

Almanac

A book or table containing a calendar of the days, weeks and months, giving statistical data on events and phenomena such as movements and phases of the moon, sunrise and sunset and tides. The derivation of the word is in doubt; the Italian is *almanacco*, the French *almanach* and the Spanish *almanaque*. An early use of the word turns up in *Opus Majus*, 1267, by Roger Bacon. Crude almanacs known as 'clogg almanacs' were in use in some parts of England till late in the 17th century. These consisted of square blocks of hard wood about 8 in long, with notches corresponding to the days of the year along the four angles. Besides the example mentioned by Roger Bacon, other early almanacs came from Peter de Dacia, about 1300, Walter de Elvendene, 1327, and John Somers, 1380.

The first known printed almanac was compiled by Pürbach and it appeared between 1450 and 1461. About 1472 there came the almanac of Regiomontanus (q.v.) probably printed in Nuremberg. In this important work the almanacs for the different months embrace three Metonic cycles, or the 57 years from 1475 to 1531 inclusive. The first English almanac printed appeared about 1497 – *The Kalendar of Shepardes*, a translation from the French. The earliest example in the United States of America came from Bradford's press in Philadelphia in 1687.

Altazimuth

An instrument for measuring the altitude and azimuth of heavenly bodies by the horizontal and vertical rotation of a telescope. Azimuth in astronomy and navigation is the angular distance, usually measured clockwise, from the south point of the horizon in astronomy or the north point in navigation to the intersection with the horizon of the vertical circle passing through a celestial body.

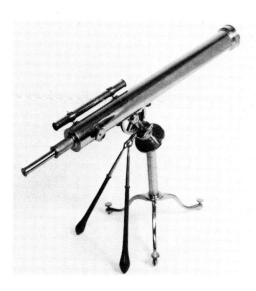

Late 19th century brass altazimuth refracting telescope by W. Harris, 47 Holborn, London, with sighting telescope, compass and pair of slow-motion adjustment rods

Alternator

A machine for producing alternating currents. Such currents change their direction of flow at regular intervals. One complete alternation or change of direction is known as a cycle or period.

Ammeter

An instrument for measuring electric current. The moving iron type is the most usual; this has a strip of soft iron that is caused to move in the magnetic field set up by the current flowing through a coil. A pointer is attached to the movable coil and registers on a scale fitted to the instrument.

Analemmatic Dial

One that is self orientating and thus re-

quires no compass; it would include an hour-arc and an azimuth dial on the same hour-plate. Analemma is a graduated scale shaped in the form of a figure eight that will indicate the daily declination of the sun.

Anaxagoras (*c.*500–*c.*428BC)

Greek philosopher who was at variance with some of his contemporaries, who had endeavoured to reduce the universe to one element such as fire or water. Anaxagoras stated that the universe was composed of an infinite number of seeds of the different kinds of matter, that in the beginning these seeds were in a state of chaos but with the arrival of intelligence a rotary impulse was given to the mass, as a result of which all cognate seeds gradually came together to form the different substances. Further, he suggested that the sun and moon were not divinities at all but lumps of red-hot stone. The sun itself was stopped from advancing beyond the tropics by a dense atmosphere that held it back. He was declared an atheist and put on trial, where he was defended by Pericles, who did secure his acquittal although he was forced to leave Athens and died at Lampsacus in Asia Minor. His most important discovery was the fact that the moon received its light from the sun; he also explained the true nature of eclipses.

Anemometer

An instrument for measuring either the velocity or the pressure of the wind. Velocity anemometers can be divided into two classes: those that require a windvane or weathercock and those that do not. The Robinson anemometer, invented in 1846 by Dr Thomas Romney Robinson of Armagh

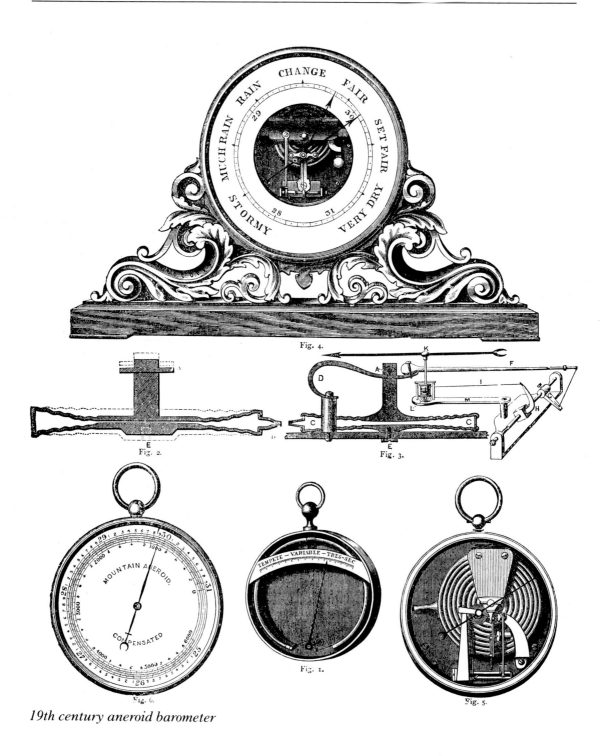

Fig. 4.

Fig. 2.

Fig. 3.

Fig. 6.

Fig. 1.

Fig. 5.

19th century aneroid barometer

Observatory is one of the best known types and does not have a directional vane. It consists of four hemispherical cups mounted on each end of a pair of horizontal arms set in the form of a cross.

Aneroid Barometer

A barometer for measuring atmospheric pressure without the use of fluids. Basically it consists of a partially evacuated metal chamber, the thin corrugated face of which is acted on by variations of the external air pressure. This movement is magnified by a system of levers which connect to a pointed reading on to a scale. The aneroid was developed towards the middle of the 19th century.

Anode

The positive electrode in an electrolytic cell.

Aperture

The opening in optical instruments; it is the portion of the diameter of an object-glass or mirror through which light can pass free from obstruction. In telescopes it is equal to the actual diameter of the cylinder of rays admitted.

Aphelion

The point in its orbit when a comet or planet is at its farthest from the sun; the point nearest to the sun is termed the perihelion. One of the most famous comets is Halley's, named after the English astronomer Edmund Halley (q.v.). This long tailed fiery traveller in space made its last nearest return to the sun in 1910 when the perihelion was 54 million miles and the aphelion about 60 times as great. Its orbital period is around 76 years.

Apianus, Petrus (1501–52)

The German scholar whose real name was Bienewitz, but this was latinized to Apianus, from the German *Biene*, a bee, which in Latin is *apis*. Born at Leisnig, he became professor of mathematics at Ingolstadt. A man of wide interests he also found fame in the fields of astronomy and geography. His most important work was the *Cosmographia*, he was also the first to publish a map of America. His son Philip (1531–89) succeeded him as professor at Ingolstadt and produced an atlas of Bavaria.

Apogee

The point in its orbit around the earth when the moon is at its greatest distance from the earth. Perigee is the point in orbit when the moon is nearest the earth. Both terms can be applied to artificial satellites in orbit round the earth.

Apothecary

The word derives from the Latin *apothecarius*, a keeper of an *apotheca*, a word used by Claudius Galen, the Greek physician born at Pergamum, Asia Minor, who was active in the 2nd century.

Apothecaries' Fluid Measure
1 minim = 0.0591 cc (about one drop)
60 minims = 1 fluid drachm = 3.55 cc
8 fluid dr = 1 fluid ounce = 28.41 cc
20 fluid oz = 1 pint = 568 cc

Apothecaries' Weights
A system of weights based on the Troy ounce and formerly used in pharmacy.
1 grain = 0.0648 gram
20 grains = 1 scruple
24 grains = 1 pennyweight

3 scruples = 1 drachm
8 drachm = 1 ounce troy = 1.1 ounce
avoirdupois

Aquinas, Thomas (1225–74)

Scholastic philosopher known as *Doctor Angelicus, Doctor Universalis*; he was born at Roccascca, the castle of his father Landuff, Count of Aquino, in the territories of Naples. Brought up in the Benedictine monastery of Monte Cassino, he went at the age of 13 to study at Naples University under the Irish monk, Petrus de Hibernia. Three years later he entered the Dominican Order, who at this time were doing their utmost to enlist the brightest scholars of the age. He adapted Aristotle's (*q.v.*) philosophy to Christian dogma, carefully distinguishing faith from reason. His philosophical theories undoubtedly influenced the scientific thinking of contemporary scholars.

Arc Lamp

A device that produces a brilliant artificial light by an arc passing between two carbon rods. To strike the arc the two rods are brought together and then rapidly separated. As the arc burns the carbon rods are vaporized and the distance between the rods is kept constant mechanically.

Archimedes (*c.*287–*c.*212 BC)

He was born at Syracuse, Sicily, the son of Pheidias, an astronomer, and he was on intimate terms with, if not related to, Hiero, King of Syracuse, and Gelo, his son. His studies were made at Alexandria; here it is likely that he first met Conon of Samos whom he admired as a mathematician and to whom later he used to show his findings and discoveries before they were published. When he returned to Syracuse he devoted himself to mathematical research. His numerous writings include papers on plane geometry – on the circle, the parabola and spirals – the sphere and cylinder, mechanics and hydrostatics. There is a story that he ran naked from the bath through the streets of Syracuse shouting '*Eureka*' (I have found it). The discovery was that the amount of water displaced by a body in a bath equals the weight of the body. Thus equal weights of different metals, such as gold and silver, when weighed in water will no longer appear equal. According to legend, the discovery was used to uncover the fraudulent practice of a goldsmith who adulterated with silver a gold crown that he was making for King Hiero.

Archimedes set his imagination to work on engines of war when the Romans were besieging Syracuse. One terrifying device was reputed to have a huge burning mirror that set light to the Roman vessels when they were in bow-shot of the walls; this is not proven as scholars such as Livy, Plutarch and Polybius make no mention of it, although it is quite conceivable that he did construct such a burning device on a smaller scale. The water screw which he invented – a device that would have been useful for the irrigation of the dry fields – was probably thought of when he was at Alexandria. Apparently he met his end when the Romans broke through and began the sack of Syracuse. He was working on a problem and refused to accompany his captor until it was finished; this so enraged the Roman soldier that he drew his sword and slew the great scholar. The action was commemorated in a mosaic believed to have come from Herculaneum, and is the only known representation of Archimedes. The Romans erected a tomb to him on which was engraved the figure of a sphere in its circumscribing cylinder.

Architectonic Sector

A sector specifically designed for use in architecture.

Architectural Clock

One that displays at the top the details of the classical pediment, with or without the columns; one or other of the Doric, Ionic or Corinthian orders might be included.

Argand Lamp

Oil lamp with a tubular wick invented by Aimé Argand, a Swiss, in 1782.

Aristarchus of Samos

The Greek astronomer flourishing around 250 BC. He is famous as having been the first to maintain that the earth moves round the sun. For this statement he was accused of impiety by the Stoic Cleanthes, just as centuries later Galileo (*q.v.*) was impugned by the theologians. His only existing work is a short treatise, with a commentary by Pappus (*q.v.*), *On the Magnitudes and Distances of the Sun and Moon*. His method for estimating the relative lunar and solar distances is geometrically correct, though the instrumental means available to him made his data erroneous. He is said to have invented two sundials, one hemispherical, the so-called *scaphion*, the other plane. A mention in Archimedes's (*q.v.*) *Arenarius*

proves that Aristarchus anticipated the discovery of the heliocentric system by Copernicus (*q.v.*). It is interesting to note that Copernicus could not have known of Aristarchus's theory because *Arenarius* was not published until after Copernicus's death.

Aristotle (384–322 BC)

The greatest of the early Greek scientists was born at Stagira, on the Strymonic Gulf, Macedonia. His father was Nicomchus, the court physician to Amyntas II, King of Macedonia and grandfather of Alexander the Great. At the age of 17 he went to Athens where he joined the school of Plato. A long association of mutual respect started between master and pupil. On Plato's death he left Athens. In 344 BC Philip of Macedon invited him to come to tutor his son Alexander on whom Aristotle was to exert a considerable influence. It is difficult to over-praise the genius of Aristotle. He stands for all time as the highest manifestation of the Greek mind with its depth of insight and never-ending curiosity. It was particularly in his scientific thought that Aristotle was distinguished. There is no branch of learning to which he has not contributed. He provided foundations for the sciences of psychology, physics and biology. The great body of his scientific thinking is incorporated in *Historia Animalum*. The *Organon*, a later collection of work, contains treatises on what he called analytics, today called logic.

Armature

The coil or coils, stationary or rotating, in an electric generator or dynamo. In a wider sense any part of an electric machine or device in which a voltage is induced by the presence of a magnetic field.

Armillary Rings

Those rings representing the Equator, ecliptic, tropic, arctic and antarctic circles etc, that, put together, make up an armillary sphere.

Armillary Sphere

A device that has the appearance of a skeletal sphere. It is composed of an assemblage of armillary rings which are designed to represent the positions of important circles of the celestial sphere. The instrument turns on its polar axis within a meridian and horizon. The principal use of armillary spheres is to demonstrate the apparent movements of the stars; some models show the motion of the planets, sun and moon.

A type of armillary sphere was said to have been in early use in China. Eratosthenes (*q.v.*) probably used a solstitial armillary for measuring the obliquity of the ecliptic. Hipparchus may have developed one with four rings. Ptolemy describes his instrument, which is of interest as an example of the armillary sphere passing into the spherical astrolabe, in Book V, Chapter 1 of the *Syntaxis*. It had a graduated circle inside which another could slide, carrying two small diametrically opposite tubes; the instrument was kept vertical by a plumbline. No great advances were made on Ptolemy's sphere until Tycho Brahe (*q.v.*) described elaborate examples passing into astrolabes in his *Astronomiae Instauratae Mechanica*.

The armillary sphere survives as a useful aid for teaching, and may be described as a skeleton celestial globe, the series of rings representing the great circles of the heavens, and revolving on an axis within a horizon. When the earth is the centre such a sphere is known as Ptolemaic; when the sun is the centre it is termed Copernican. There are survivors from medieval times until around the late 17th century. Static, manual and mechanical spheres were made. Armillary spheres occur in a number of old sculptures, paintings and engravings; from such sources it is possible to gather that they were made to be suspended, placed on the ground or a table, held by a short handle or rested on a stand.

Ptolemaic armillary sphere by Francis Lagshinus, 1602

English mechanical armillary sphere, 18th century

Ashurbanipal (ruled 669–626 BC)

King of Assyria, the grandson of Sennacherib, the most powerful and by repute brutal monarch of his time, he ruled Babylon, Egypt, Persia and Syria. Under him art, literature and science reached a high level as shown when his library at Nineveh was excavated by Sir Austen Henry Layard during the years 1845–47; it held some 22,000 clay tablets inscribed in cuneiform with Assyrian, Babylonian and Sumerian historical, literary, religious and scientific records. Clues from these guided Sir Leonard Woolley to rediscover sites of Mesopotamia's first civilization; he worked at Ur of the Chaldees, the birthplace of Abraham, from 1922 to 1934.

Aspirator

An apparatus so constructed as to allow the drawing of a current of air or other gas through a liquid.

Astatic

Not static, unstable, having no tendency to adopt any particular orientation or position.

Astatic Coils

Coils used with very sensitive electrical instruments, they are so arranged that the magnetic force produced when a current is passed through them and the electro-motive force induced in them by an external magnetic field are zero.

Astatic Galvanometer

A type of moving magnet galvanometer which has two mutually compensating magnets arranged parallel but in opposition at the centres of two coils oppositely wound.

As the resulting magnetic moment is zero the instrument is not affected by the earth's magnetic field.

Asterism

A cluster of stars or a constellation. Derived from the Greek *asterismos*.

Asteroids

Minor planets that move around the sun, mainly between the orbits of Jupiter and Mars. The largest of them is Ceres which is less than 500 miles in diameter compared with Mars which is 4,332 and Jupiter at 87,380. More than 600 other asteroids, including Pallas, Juno and Vesta, have been discovered. In 1891 Dr Max Wolf (*q.v.*) of Heidelberg developed a method for using photography in tracking these bodies down. He worked out that if a photograph was taken through an equatorial telescope – moving by machinery so as to keep the stars at which it is pointed always exactly in the field of view during the apparent movement across the sky – the images of the stars would be sharply defined points. If, however, there is an asteroid in the same field of view, its image will appear as a white streak because the body has a comparatively rapid motion and will, during the period of exposure, move sufficiently to mark its passage.

Astrarium

An elaborate astronomical clock that became popular during the early part of the 15th century, mainly in churches and public buildings. The finest models were magnificent examples of craftsmanship and cost limited their ownership to kings, princes, nobles and church dignitaries. One of the most outstanding models, by Giovanni de'Dondi, was also one of the earliest. De'Dondi worked on it from 1348, probably assisted by his father Jacopo who was the municipal physician of Chioggia. The idea first came to Giovanni from the *Theorica Planetarum* of Giovanni Campano da Novara or John Campanus, the text being one of the earliest (*c.*1270) and most important works on the equatorium in medieval Europe. The instrument contained, amongst other parts, an astrolabe, dials for mean time, sunrise, sunset, Mars, Venus, clock movement and indicators for the fixed and movable feasts of the Church. De'Dondi forged all its parts of brass and copper and worked on it for about 16 years. Evidence of the quality of the astrarium appears in a letter written on 11 July 1388 to De'Dondi by his friend Giovanni Manzini who thought that it was a thing:

> ...full of artifice, worked on and perfected by your hands and carved with a skill never attained by the expert hand of any craftsman ... I conclude that there was never invented an artifice so excellent and marvelous and such supreme genius.

The patience of the man is shown by the fact that the 107 wheels and pinions had to be hand finished with a file and, as screws were not used, the whole had to be assembled with more than 300 taper pins and wedges.

Astro-Compass

A navigational instrument for determining direction relative to the stars. It gives directional bearings from the centre of the earth to a particular star. It is unaffected by the errors that magnetic and gyro compasses are prone to and can be used to check the accuracy of such instruments.

Astrolabe

Instrument used for taking the altitudes of the sun, moon and stars. There were two types: planispheric and spherical. The earliest forms were 'armillae' and spherical. Their development can be traced from Eratosthenes (*q.v.*) to Tycho Brahe (*q.v.*) and a leading part among the ancient astronomers was played by Hipparchus (*q.v.*). The complex astrolabe was evolved and

Persian brass astrolabe, signed Ali ibn Sadiaz

remained one of the chief observation aids throughout the 15th, 16th and 17th centuries; small models were used by travellers and scholars not only for astronomical, but also for astrological and topographical, purposes. Examples could include the ability not only to take the angle of heavenly bodies but also to calculate latitude and points of the compass, and they could also be used to find the heights of mountains. Up until the middle of the 18th century they were used by sailors as a navigational instrument to determine latitude and time.

Mid-17th century North African astrolabe

Astrology

What some consider the ancient art and others the pseudo-science of divining the fate and future of mortals by seeming to read indications in the stars. For long ages astronomy and astrology, which might be called astromancy, were confused. Some made a distinction between 'natural astrology', which predicts the motions of the heavenly bodies, eclipses etc, and 'judicial astrology' which studies the influence of the stars on the destiny of mortal man. Isidore of Seville (d.636) was one of the first to truly make a clear distinction between astrology and astronomy although the scholars studying astronomy did not really manage to rid the science of the superstitious ways of astrology until the 16th century.

The pursuit of astrology and the confusion with astronomy started with the Babylonians who spread their findings to the Middle East in general and from there to Greece in the 4th century BC. The doctrine reached Rome before the Christian era. From Rome it spread throughout the world, at times causing concern when cases of what might be termed 'astrological politics' occurred or 'heaven-sent' rulers appeared. Both Napoleon and Wallenstein believed in their stars.

Astronomical Clock

One that will point to or show by moving dials the phases of the moon, the rising and setting of the sun, also with some examples the principal planets, as well as telling the time in minutes and hours. There is an impressive example at Beauvais by A. Verité.

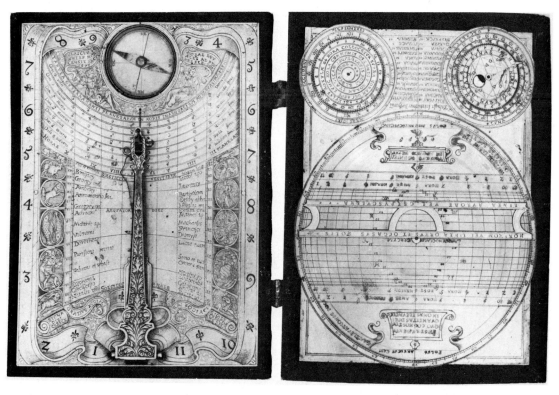

A rare astronomical compendium of diptych form (by Ulrich Schniep), one leaf signed within a cartouche: VLRICVS SCHNIEP DE MONACO FECIT

Astronomical Compendium

It could be a combined set of a number of instruments or a complex device capable of serving many purposes; it was developed rather than invented during the 16th and 17th centuries. The compendium could be circular, hexagonal, octagonal or rectangular boxes, often beautifully decorated with gilt-brass fittings; some had the appearance of books. Instruments contained could be several astronomical devices such as sundials, nocturnals, wind roses, compasses, lunar volvelles and often latitude tables, also possibly an astrolabe, quadrant and a set of drawing instruments. One notable maker was Christoph Schissler, active in Augsburg in 1557.

Astronomical Ring

One developed from the armillary sphere which can help with telling the time. It is very rare.

Astronomy

The scientific study of the individual celestial bodies, but not the earth, and of the universe as a whole; in other words, the investigation of all matter of the universe outside the limit of the earth's atmosphere. It is the most ancient of the sciences, because before the time of experiments, it was the branch of knowledge which could be most easily systematized, while the relations of its phenomena to day and night, times and seasons could be easily understood in their simplest forms by most people.

It can be divided broadly into two parts: astronomy proper or Astrometry, which deals with the movement, dimensions and relation of the heavenly bodies and Astrophysics which studies their physical constitution. Practical astronomy in its broadest and most literal sense deals with the instruments by which we observe the heavenly bodies, the basic principles of their use and the methods by which these principles are applied in practice.

It is difficult to date accurately the first studies in astronomy but certainly the equinoxes and solstices were being determined in China during the third millennium BC. Much of the progress is charted by the invention and development of the instruments that enabled problems to be solved and the eye and mind of man to reach further and further out into the great expanses of the unknown. At the moment another 'largest telescope in the world' is being planned. British astronomers are investigating a 'monster' that would have ten times the power of the biggest existing telescope. This gigantic instrument will have six 25 ft mirrors which, when combined, will provide the power of a telescope with a 60 ft mirror; the whole thing may be sited on top of a 7,800 ft mountain at Las Palmas in the Canary Islands.

Augsburg Dial
Small universal equinoctial dial, generally in a finely finished and possibly decorated case that would fit in the pocket. They are associated with the instrument makers of Augsburg and were also popular in France in a somewhat different form.

Automatic Winding
A timepiece so constructed that the motion of the wearer will keep the spring wound and operating.

Aux
In Ptolemaic theory the direction, as seen from the earth, of the centre of the deferent (a circle centred on the earth around which the centre of the epicycle was thought to move) of the sun, moon or a planet. When the sun, moon or the centre of a planetary epicycle is said to be in the aux direction it is at the apogee of its deferent and thus at the farthest possible distance it could be from the earth.

Azimuth Dial
One with a pin-gnomon dial that has concentric hour-scales graduated in the solar azimuth angles for every month of the year.

B

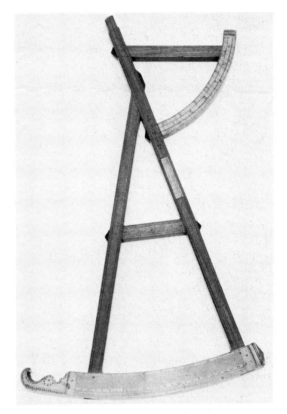

18th century mahogany back staff, scales engraved on boxwood arcs

Back Staff

A device similar to a cross staff: it was formerly used for taking the altitude of the heavenly bodies.

Baily, Francis (1774–1844)

English astronomer born at Newbury, Berkshire who, after serving his apprenticeship in London, went to America and travelled the unexplored parts of the west. On his return to London in 1798 he joined the Stock Exchange and made himself a considerable fortune. He retired and fitted up an observatory at his house in Tavistock Place, London and devoted himself to astronomical study. He was one of the founding members and later president of the Astronomical Society. He re-edited the star catalogues of Ptolemy (*q.v.*), Tycho Brahe (*q.v.*), Halley (*q.v.*), Flamsteed (*q.v.*) and others. In his observations he was the first to note the strange knobs of light on the rim of the sun during a total eclipse, the phenomenon has since been known as Baily's Heads. His estimate of the weight of the earth has been accepted as close to the approximate truth. He was treasurer of the Royal Society and one of the founders of the Royal Geographical Society.

Balance

A term originally used for the ordinary beam balance or weighing machine with two scale pans; in the simplest form it is a horizontal beam pivoted at its centre. The name can also include other apparatus for

Trade label from the box for a beam balance, 18th century

comparing and measuring weights and forces. Besides the beam and spring balances there is the 'torsion' balance with which forces are measured or compared by the twisting moment on a wire. This might

be used in experiments on gravitation, electrostatics or magnetic fields.

Balance Staff
The arbor or shaft that carries the balance wheel in a clock.

Balance Wheel
A wheel oscillating against the hairspring of a clock or watch to regulate the movement; it was one of the first forms of such control which enabled the timepiece to be as accurate as a pendulum clock. Christiaan Huygens (*q.v.*) in 1675 announced he had made a spiral spring which could be fitted to the balance wheel; round about the same time Robert Hooke (*q.v.*) claimed that he had already produced such a device.

Ballistic Galvanometer
One for measuring the exact quantity of electricity passing through a circuit due to a momentary current. It is possible to use any galvanometer to do this if its period of oscillation will amply cover the time that the momentary current flows.

Brass inspector's beam scale by de Grave, London

Balloon Clock

A design for clocks that became popular in the late 18th century and early 19th. They tended to have comparatively large round faces, drawn-in waists and then spread out again at the base, chiefly found as a model for the mantelpiece or shelf.

Banjo Barometer

A instrument in which the case is made in the form of a banjo; the bulbous end, displaying the dial, is at the bottom.

Banjo Clock

An American pendulum wall clock in this shape. The pendulum door is often decorated with *églomisé* panels, favourite subjects being views of Lake Erie and Mount Vernon. The first example was probably made by Simon Willard before 1800; the design was patented by him in 1802.

Barograph

An elaborate form of aneroid barometer in which the diaphragm movement is connected to a pen that records these movements on a roll of paper kept in motion mechanically.

Barometer

An instrument by which the pressure or weight of the atmosphere can be measured. The normal mercury barometer has a glass tube that is hermetically sealed at the top and is then filled with mercury. If the tube is placed with its open end in a basin of mercury, it is termed a 'cistern barometer' and the atmospheric pressure is measured by the difference of the heights of the mercury in the tube and the cistern. With a

'siphon barometer' the cistern is not used, the tube is bent round in a 'U' shape and the reading is taken between the levels in the two arms (*see* Aneroid Barometer). The Fortin is a refinement to the cistern barometer in which the upper part of the cistern is formed of a glass cylinder through which the level of the mercury can be observed and the bottom of the cylinder is sealed with a soft flexible leather bag supported by a small metal plate against which a screw works, this allows for alterations in the mercury level to adjust it to the zero on the scale.

Mahogany angle barometer with thermometer framing a perpetual calendar by F. Watkins (1746–c.1784), London

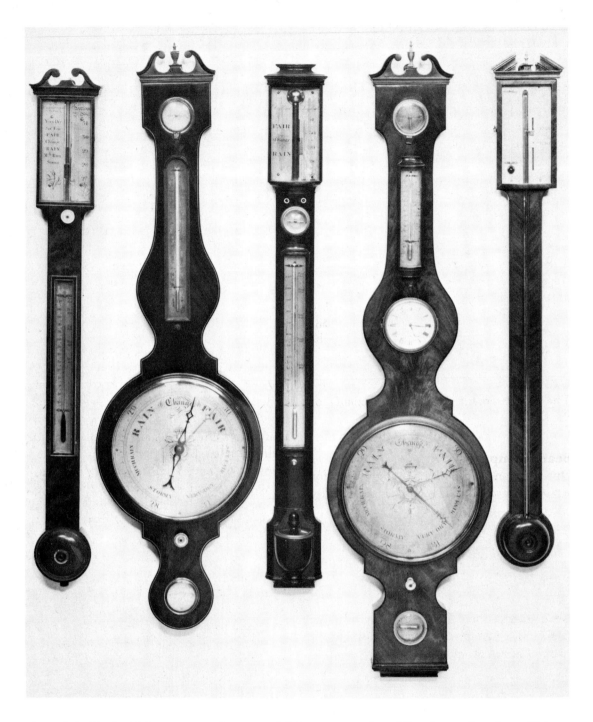

From left: Early 19th century mahogany stick barometer, signed Berringer, London. *19th century mahogany wheel barometer by Lewis Arnaboldi, 7 Watton Place, Blackfriars, London. George III bow-front stick barometer, signed* Dollond, London. *19th century mahogany wheel barometer with level, signed G. Rofsi, Norwich; timepiece signed* Beefield, London; *includes thermometer and hygrometer. Georgian mahogany stick barometer by Nathan Dawson*

Barrel

A vessel of cylindrical shape. The term can be applied to a number of cylindrical objects: the barrel of a telescope or microscope, the drum round which the wire or chain of a crane is wound, a capstan and the cylinder studded with pins in a barrel organ or musical box.

Basket Top

A distinctive pierced metal dome on bracket clocks from the last part of the 17th century to the early years of the 18th. One variation was the double basket top with two superimposed domes, another was to have the dome made from a hard wood.

Bathymetry

The measurement of depth, especially of the sea.

Beam Compass

One that can be used by a draughtsman for circles or arcs of large radii. It generally consists of square or hexagonal lengths of metal rods that will fit together. One end is secured to a pin on a base whilst the pencil or ink points are in holders that will slide along the jointed bars.

Beckmann Thermometer

A sensitive instrument for measuring very small changes in temperature.

Beehive Clock

A small shelf clock originating in Connecticut. The name is derived from the fact that there is some resemblance to the old kind of beehive; it dates from about 1850.

Behaim, Martin (c.1436–1507)

Navigator and geographer extraordinary, was born at Nuremberg some say in 1436, although Ghillany gives the date as 1459. A confirmed merchant, he was drawn to Portugal for participation in the Flanders trade and whilst there acquired a scientific reputation at the court of John II. Whether or not he was actually a pupil of the astronomer Regiomontanus (q.v.) is not certain, but rumour of this helped his appointment as a member of a council formed by King John to further the study of navigation. He has been credited with the introduction of the cross-staff to Portugal; he also made astrolabe models in brass that were lighter in construction and more easily handled than some of the earlier examples which had, in a number of cases, been made of wood, and it is likely that he helped prepare more advanced navigation tables. Behaim claimed to have sailed with Diego Cão when he went on his second expedition to West Africa; carping critics suggest that this was not so and state that Behaim only went as far as the Bight of Benin and possibly with José Visinho, the astronomer. On his return from the voyage a grateful King John knighted him. He is probably best recalled as the maker of the oldest surviving globe. In 1492 he returned to his native Nuremberg and, working with the painter Glockenthorn, produced a segmented map as the basis for the construction of his terrestrial globe. The globe leans heavily on the Ptolemaic ideas of geography and as a work for scientific study is amusing rather than helpful. Details of West Africa are incorrect, the Cape Verde Islands are hundreds of miles out of place and the Atlantic is liberally peppered with strange and wonderful islands.

Bell, Alexander Graham (1847–1922)

American inventor and physicist, the son of Alexander Melville Bell, the educationist and specialist in speech defects. He was born in Edinburgh, and after periods at Edinburgh and London Universities he left for Canada with his father. In 1872 he was appointed professor of vocal physiology at Boston University. Four years later he exhibited a device that transmitted sound by electricity, the development of which owed much to his research into speech mechanics whilst he was training teachers for the deaf. After further development and refining this apparatus became the ancestor of today's telephone. He also invented a gadget called the photophone which could transmit sound by variations in a beam of light, an electric probe for detecting bullets in the body and the grooved wax cylinder with a spiral sound track for Thomas Edison's (*q.v.*) phonograph.

Bell Top

A type of clock with the lower portion of the top of its case shaped like the bell of a turret clock with concave sides.

Bezel

Metal framing of a watch glass or clock. Also, in gem cutting, the rim which secures a jewel in its setting.

Binocular

An apparatus through which objects are viewed with both eyes. The first binocular telescope, which was really two telescopes

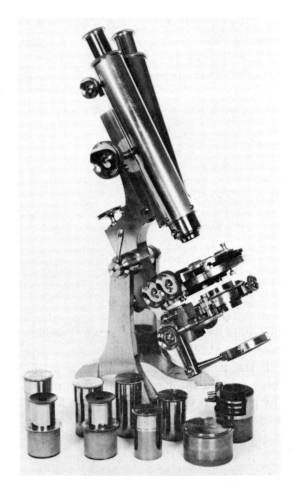

Ross brass binocular microscope with rack-and-pinion and milled screw focusing, parabolic condenser and accessory lenses

placed side by side, was constructed in 1608 by Johann Lippershey. In 1645 the Capuchin Antonius Maria Schyrläus de Rheita (1597–1660) described the making of double terrestrial telescopes. Another contemporary Capuchin, Chérubin d'Orléans, made large double telescopes of the Dutch type with high magnification for use in war, and smaller instruments of lower magnification which had a mechanism for adjusting the interval between eyepieces to suit the eyes of the observer. No further important work was done on this until the re-invention of the Dutch binocular telescope around 1823, much of the development being credited to Johann Friedrich Voigtländer (1779–1850).

The first binocular microscope was constructed by Father Chérubin. His instrument consisted of two inverting systems and, in consequence, gave a totally false impression of depth, a situation described by Charles Wheatstone (q.v.) as a 'pseudoscopic impression'. Again it was the 19th century before further steps in this direction were taken. In 1853 J. L. Riddell (1807–67) constructed a binocular microscope which still contained something of the 'pseudoscope'. F. H. Wenham, who was working with the firm of Ross, designed a prism which formed the basis of his instrument. The prism was so set as to cover one half of the objective lens, and deflected the rays into the second ocular tube which was fixed at an angle to the other ocular tube which was vertical.

Bion, Nicolas (active around 1700)

French instrument maker; a fine craftsman, who often worked in silver. He completed a number of objects for the eastern Mediterranean market, probably for the Turks who controlled an important fleet.

Biquintile

The aspect of two planets which are distant from each other by twice the fifth part of a great circle. This was one of the new aspects introduced by Kepler (q.v.).

Bird, John (1709–76)

London instrument maker who was not only concerned with the construction of devices but also actively researched the problems that they were intended to solve.

Bloud Dial

A form of diptych dial whose design is associated with Charles Bloud and other instrument makers in Dieppe around 1660. The principal feature is a magnetic azimuth dial; often also included were equinoctial, polar and string-gnomon dials. The dials were often made of ivory.

Bob

The weight at the base of a pendulum rod which could generally be screw adjusted to control the loss or gain of the clock. Early ones were shaped something like a pear or almost spherical, later they became almost universally disc shaped.

Bolometer

Very sensitive instrument used for measuring heat radiations. To determine the intensity of radiations along the spectrum band another type termed a spectrum bolometer can be used.

Bone, William Arthur

English inventor born in 1871 who did research on combustion at Owens College, Manchester and later at Leeds University between 1905 and 1912. This resulted in the discovery of a method of flameless combustion named after him and his chief fellow researcher, C. D. McCourt – the Bonecourt system. This invention became a revolutionary factor in all smelting and allied industries. In 1912, he was appointed professor of fuel and refractory materials at the Imperial College of Science, London.

Borelli, Giovanni Alfonso (1608–79)

Italian astronomer, biologist and mathematician, born in Naples. He was appointed professor of mathematics at Messina in 1649 and at Pisa in 1656. In a letter, *Del movimento della cometa apparsa il mese di decembre 1664*, published in 1665 under the pseudonym Pier Maria Mutoli, he was the first to suggest the idea of a parabolic path. This was followed by *Theorica mediceorum planetarum ex causis physicis deducta*, published in Florence in 1666, which outlined his thinking on the attraction of the satellites of Jupiter.

Boscovich, Roger Joseph (*c*.1711–87)

Italian mathematician and naturalist born at Ragusa, Dalmatia. He was one of the earliest foreign scholars to adopt Newton's (*q.v.*) theory of gravitation. When he was 15, he entered the Society of Jesus and on completing his noviciate he studied mathematics and physics at the Collegium Romanum; so brilliant were his results that in 1740 he was appointed professor of mathematics at the college. Subjects that he published papers on included the transit of Mercury, the aurora borealis, the figure of the earth, observation of the fixed stars, inequalities in terrestrial gravitation, application of mathematics to the theory of the telescope, the theory of comets, the tides, the double refraction micrometer and spherical trigonometry. In 1758 he published in Vienna his most famous work *Theoria philosophiae naturalis redacta ad unicam legem virium in natura existentium*, which outlined his theory that the atom has continuity of existence in time and space.

Bose, Sir Jagadis Chandra

Indian scientist, born in 1858, who invented the crescograph, an instrument, which by enormous magnification of the natural movements of plants, makes it possible to record their life-growth. He was president of the Indian Science Congress in 1927.

Boyle, Robert (1627–91)

Praised by many as one of the strongest influences on the progress of science, he was the son of the 1st Earl of Cork and was born at Lismore Castle in the province of Munster, Ireland. While still a child he spoke French and Latin and went to Eton when only eight years old. He visited Italy in 1641 and studied the work of Galileo (*q.v.*). Back in London in 1657 he read an account of Otto von Guericke's air pump. He set to work with the help of Robert Hooke (*q.v.*) to improve on the design of this device until in 1659 they had completed a 'pneumatical engine' which enabled him to carry out a series of experiments on the properties of air; the results were published under the title *New Experiments Physico-Mechanical touching the spring of air and its effects*. His other inventions and research included thermometers, barometers, freezing mixtures, gas laws, chemical elements and electricity. His other publications included *New Experiments and Observations upon Cold, Cosmical Qualities of Things, Hydrostatical Paradoxes* and *Celestial Magnets*. Boyle put into practice the theories that Bacon evolved in the *Novum Organum*, and although for many years he refused to read Bacon or even Descartes unless he was influenced by them, it is as the interpreter of their works that he is of the greatest importance. Towards the end of his life he spent much serious study on theology and he bore the cost of translating the Bible for readers in India, Ireland, Malaya, Turkey and Wales.

Bracket Clock

One designed to stand on a shelf or special bracket. They had short pendulums and could often be of considerable size, comparable to the dial diameter of the largest grandfather or longcase clocks.

Brahe, Tycho (1546–1601)

The Danish astronomer was born at the family seat of Knudstrup in Skåne, then a Danish province. He was adopted by his uncle Jörgen Brahe who sent him to Copenhagen to study philosophy and rhetoric. An eclipse which happened at the predicted time, 1560 August 21, inspired him to look upon astronomy as 'something divine'. He procured a copy of *Ephemerides* by Johann Stadius and the works of Ptolemy (*q.v.*) in Latin and began to gain some insight into the theory of the planets. Although entered as a law student at the University of Leipzig, he secretly worked hard on celestial studies and started continuous observations with a globe, a pair of compasses and a cross staff. He left Leipzig and went first to Wittenberg and then to Rostock where in 1556 he lost his nose in a duel and had a false one made from a copper alloy. When he returned to Denmark he was encouraged by his maternal uncle, Steno Belle, to fit out a laboratory at his castle of Herritzvad, near Knudstrup. Here it was in November 1572 he caught sight of a 'new star', Cassiopeia; he carefully measured its position and published a paper *De Nová Stellâ* in 1573. In 1576 Frederick II, the King of Denmark, bestowed on him for life the island of Hven in the Sound near Elsinore; here Tycho built Uraniborg, his Castle of the Heavens, which, on its completion in 1580, became a mecca for astronomers from all over Europe. The instruments were the finest obtainable and he himself produced many that were refinements of those in existence. He worked to correct many astronomical theories but latterly he was hindered, for with the death of Frederick II, the new monarch, Christian IV, was less tolerant of Tycho's insubordinate and arrogant manners and withdrew the benefits he had received. In 1597 he returned to Rostock and restarted work at the castle of Wandsbeck; later moving to Prague and coming under the protection of the Emperor Rudolph II who gave him a handsome pension and the use of the castle of Benatky. His instruments arrived from Hven and he was joined in 1600 by Johann Kepler (*q.v.*). He died in 1601 before he could achieve his full potential.

His principal work was entitled *Astronomiae Instauratae Progymnasmata*, this was edited by Kepler and published in Prague in 1602–3. It included much on the subjects closest to his heart – motions of the sun and moon, it gave the positions of 777 fixed stars (this number was increased by Kepler in 1627 to 1,005 in the Rudolphine Tables). The second volume contained, under the heading *De Mundi Aetherei recentioribus Phaenomenis*, the comet of 1577, his plan of the cosmos, also his belief that the earth retained its immobility and that the sun revolved round it.

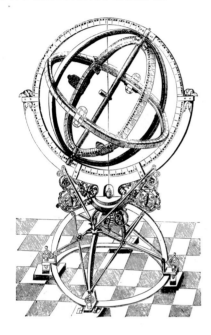

An equatorial used by Tacho Brahe in his astronomical calculations

Bramah, Joseph (1748–1814)

English engineer and inventor, born at Stainborough, Yorkshire, the son of a farmer. He worked first as a cabinet-maker in London where he then started business on his own account. In 1748 he patented the ingenious lock that bears his name; he was indebted to Henry Maudslay (*q.v.*), one of his workmen, who assisted him in designing a machine for the manufacture of his locks. Other inventions included a mechanism for water-closets, his hydraulic press, a machine devised for the Bank of England for numerical printing, especially adapted for bank notes, paper making machinery, and he suggested the possibility of screw propulsion for ships.

Brewster, Sir David (1781–1868)

Scottish natural philosopher, born at Jedburgh where his father was rector of the grammer school. At the age of 12 he was sent to Edinburgh University. He had already become an admirer of the self-taught astronomer and mathematician James Veitch of Inchbonny, who was a skilled maker of telescopes. Although he finished his studies and became qualified to preach, his mind was on other matters. In 1799 he was led to study light by a fellow student, Henry Brougham. The most important subjects of his research were the discovery of the polarising structure induced by heat and pressure, the laws of polarisation by reflection and refraction, the discovery of crystals with two axes of double refraction, the laws of metallic reflection and experiments on the absorption of light. He rediscovered the kaleidoscope in 1815, and developed the stereoscope, particularly the lenticular, which was very largely his work. He followed up Fresnel's work on lighthouses and introduced the holophotal system of illumination by which the whole light of the lamp, whether reflected or refracted, could be used. In 1855 he issued *Memoirs of the Life, Writings and Discoveries of Sir Isaac Newton* (*q.v.*), the result of some 20 years' study of original manuscripts and other available sources.

Briggs, Henry (1561–1630)

English mathematician, born at Warley Wood in Yorkshire. He graduated at St John's College, Cambridge in 1581. In 1596 he became the first professor of geometry in Gresham House (afterwards College), London. In his lectures he proposed the alteration of the scale of logarithms from the hyperbolic form which John Napier (*q.v.*) had given it, to that in which unity is assumed as the logarithm of the ratio of ten to one. The adoption of logarithms for astronomical purposes was largely due to his work, which was accepted throughout Europe.

Bright, Sir Charles Tilston (1832–88)

English telegraph engineer, born at Wanstead, Essex; at the age of 15 he became a clerk in the Electric Telegraph Company. His talent for electrical engineering soon brought him promotion to engineer with the Magnetic Telegraph Company and in this capacity he was responsible for the laying of cables in Britain, the first cable between Britain and Ireland and subsequently the first Atlantic cable. Working with Josiah Latimer he invented new methods of insulating submarine telegraph cables.

Brinkley, John (1763–1835)

English astronomer born at Woodbridge, Suffolk. He was appointed professor of astronomy in the University of Dublin in 1792 and later Astronomer Royal for Ireland, and director of the College Observatory at Dunsink, near Dublin. Here he carried out observations until he was made Bishop of Cloyne in 1826. His principal written work was *Elements of Astronomy*.

Bunsen, Robert Wilhelm Von (1811–99)

German chemist born at Göttingen, where his father, Christian Bunsen, was the chief librarian and professor of modern philology at the University. In 1852 Robert was appointed to the chair of chemistry at Heidelberg where he was to spend the rest of his working life. In his early years there he was concerned with research into arsenic and nearly killed himself by poisoning and he also lost the sight of one eye through an uncontrolled explosion. In 1841 he invented the carbon-zinc electric cell and later, to measure the strength of light, he developed the grease-spot photometer. Between 1855 and 1863 he published, with Roscoe (*q.v.*), papers on photochemical measurements. His most famous invention, the Bunsen burner, came in 1855; it was quite a simple device for burning ordinary coal gas with a hot smokeless flame. It was basically an upright tube with an adjustable sleeve towards the bottom that regulated the amount of air being mixed with the coal gas. He also invented an ice-calorimeter in 1870, a filter pump in 1868 and a vapour calorimeter in 1887. Working with G. R. Kirchhoff, he made studies in depth of the light emitted by compounds heated to incandescence; his work on spectrum analysis brought enlightenment to astronomers and chemists. At the start he was able to isolate two new elements of the alkali group, caesium and rubidium. He was a man of outstanding perseverance and patience, for it is recorded that in this particular piece of research he evaporated some 40 tons of Dürkheim water to obtain just 17 grammes of the mixed chlorides of the two substances. He was a highly respected teacher and students flocked to his lectures; practical work to him was all important and he held that truth alone was the end of scientific research.

Butterfield, Michael (active 1660)

Instrument maker from Paris, to a degree specializing in the smaller devices and working to high standards. He produced a range of instruments for the collector as well as for popular demand.

Late 17th century French Butterfield dial, signed Butterfield à Paris, *the gnomon supported by a bird; engraved with Continental towns and cities and their latitudes*

Byrgius, Justus (1552–1632)

A Swiss who invented a proportional compass which was of great assistance to draughtsmen and workers in geometry; he also devised a system of logarithms.

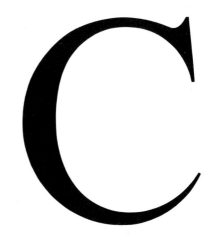

Cabot, John (c.1450–c.1500)

Navigator born in Genoa, then lived in Venice; he came to England about 1484. In 1496 letters of patent were issued to him by Henry VII which in part read:

... well-beloved John Cabot, citizen of Venice, to Lewis, Sebastian and Santius, sonnes of the said John, full and free authority, leave and power upon theyr own proper costs and charges, to seeke out, discover and finde whatsoever isles, countries, regions or provinces of the heathen and infidels, which before this time have been unknown to all Christians.

It was made clear that any merchandise arising from all this should be allowed to enter the port of Bristol free but that one fifth of all gains should go to the King. Inspired by this, John Cabot and 18 others set sail from Bristol in the *Mathew* on 2 May 1497. After rounding Ireland they pitted themselves and their small craft against the vagaries of the Atlantic for 52 days until at 5 am on Saturday 24 June they reached Cape Breton Island and Cabot was convinced that he had landed on the north east coast of Asia – the source of all those rich silks and precious stones that had led them on. He returned to Bristol and with fresh letters of patent from Henry set out again for the promised land of the Orient, this time with ten ships and 300 men. Early in June 1498 he made a landing in Greenland which he named Labrador (later the name was transferred to a coastal area of Canada); sailing northwards into the misty cold of those lonely waters eventually the crews mutinied and after more vicissitudes he returned to Bristol.

Cabot, Sebastian (*c.*1474–1557)
Son of John Cabot (*q.v.*) with whom he sailed on the voyage to Cape Breton. After this facts about his life are indistinct until in 1512 there is a mention of his making maps of Gascony and Guienne for Henry VIII and also various voyages across the Atlantic including a landing in Brazil. He returned to England where he founded the Merchant Adventurers. His map of the world was published in 1544.

Calculating Machines
Instruments for the mechanical performance of numerical calculations can be divided broadly into five groups: plain addition devices, the first of which was invented by Blaise Pascal in 1642; addition with modifications for multiplication, introduced by G. W. Leibnitz in 1671; true multiplication machines by Léon Bollés in 1888; differ-

ence machines by Johann Helfrich von Müller in 1786 and Charles Babbage in 1822; he also introduced analytical devices in 1834. The best account of the development of these machines was given by Mehmke in *Encyclopädie der mathematischen Wissenschaften*.

19th century brass Lords patent calculator by W. Wilson, London. Silvered dial engraved with three scales; pointers measuring linen lea, worsted hank, linen bundle and cotton and silk hank.

Calendar Clock
One which indicates the hour, day, month and year. These timepieces have been popular for centuries. Probably the earliest known is the clock by Giovanni de'Dondi, of Padua; this has an extra refinement

which shows the amount of daylight to be expected each day.

Callan, Father Nicholas (1799–1864)

Priest-scientist who was born at Darver, Co Louth, Ireland. His studies at St Patrick's College, Maynooth went far beyond those of the average student and before he was 30 he was making significant discoveries of his own and predicting the coming of later technological developments. He remained at Maynooth where he became professor of natural philosophy, a post he held until his death. Records show that by the 1830s Father Callan had become Ireland's leading figure in the field of electromagnetics. Devices on which he worked included the Callan coil, a mechanical interrupter, the so-called Maynooth battery, a repeater, a self-induced dynamo and a galvanizing system for which he received a patent. The Grant to Father Callan was made in 1853 and reads in part:

Whereas Nicholas Callan of the RC College of Maynooth in the County of Kildare Ireland — hath by his petition humbly represented unto us that he is in possession of an invention for 'a means of protecting iron of every kind against the action of the weather and of various corroding substances so that iron thus protected will answer for roofing, cisterns, baths, gutters, pipes, window frames, telegraph wires, for marine and various other purposes' which the petitioner believes will be of great public utility that he is the first and true inventor thereof and that the same is not in use by any other person or persons to the best of his knowledge and belief . . .

Father Callan's cast iron battery. This was marketed by a London instrument maker as the Maynooth battery, c.1839.

Callipers

Instruments for measuring internal or external dimensions of tubes or from point to point on the flat. They consist of two legs hinged together. There is sometimes an arc which, by the aid of a screw attachment, can clamp and hold the measurement between the legs.

Calorescence

A name invented by John Tyndall (*q.v.*) for an optical phenomenon, the essential feature of which is the conversion of rays belonging to the dark infra-red portion of the spectrum into the more refrangible visible rays; that is, heat rays into rays of light.

Calorimeter

Instrument for the measurement of heat energy, to find specific heat capacities and calorific values.

Camera Lucida

An optical instrument to aid drawing in perspective without establishing construction lines such as the horizon line and viewing plane and without placing the vanishing points. The early model was invented by Dr William Hyde Wollaston (*q.v.*) at the beginning of the 19th century. The basic theory was that by closing one eye and looking vertically downwards with the other through a piece of plain glass held at an angle of 45 degrees to the horizon, images of the scene in front could be noted by reflection on the surface of the glass whilst at the same time the eye could see through the glass, and a pencil or other drawing instrument could be guided along the seeming lines of the composition. More sophisticated instruments with which the eye could view through a prism and glasses of varying degrees of magnification could be slipped into the path of vision for increasing the size of the image to be drawn followed.

Camera Obscura

A darkened chamber often large enough to accommodate one or more observers, the top of which contained a box or lantern with a convex lens and a sloping mirror, or a prism which combined the purpose of the lens and mirror. With this it was possible to project images of objects outside the device on to a flat surface inside. It is likely that the instrument was invented by the 16th-century Neapolitan, Giovanni Battista della Porta, although simple versions of the principle had been in use for some time for observing eclipses. There are indications that the Arabian philosopher Alhazen investigated the idea around 1038. Roger Bacon was certainly acquainted with the principle of the camera obscura. In 1437 the architect Leon Battista Alberti was making progress with the idea and Leonardo da Vinci (*q.v.*) also had such thoughts. Kepler (*q.v.*) used the device for solar observation in 1600 and in his *Ad Vitellionem Paralipomena* of 1604 poses the problems of the passages of light through small apertures, and the purpose of the simple darkened chamber; he was the first to describe an instrument with a sight and paper screen for observing the diameters of the sun and the moon in a dark room. In 1611 Johann Fabricius describes how he and his father could observe the sun's spots just as well with an obscura as with a telescope. In Robert Boyle's (*q.v.*) essay *On the Systematic or Cosmical Qualities of Things*, which was written about 1670, he mentions a kind of box camera with a lens that could be used to view landscapes. Johann Zahn tells of similar ideas in his *Oculus Artificialis Teledioptricus*. William Molyneux of Dublin in his *Dioptrica Nova* in 1692 treats the device mathematically; and Sir Isaac Newton (*q.v.*) in his *Opticks* (1704) goes into the principle of the obscura.

Candle Clock

Primitive and unreliable timing device relying on the burning speed of a wax or tallow candle; inaccuracies were almost certain because of changes in the diameter of the candle and the different burning speeds of the wax or tallow.

Canes Venatici

A northern hemisphere constellation named by Hevelius (*q.v.*) in 1690. Also referred to as the Hounds or Greyhounds.

Capacitator

An electric device for storing static electricity, often consisting of two conducting surfaces separated by a dielectric.

Carat

A small weight originally in the form of a seed used for weighing gems; also, a measure for establishing the fineness of gold.

Cardan, Girolamo (1501–76)

The name is also spelt Cardano, Hieronimo and Cardano, Geronimo. Italian mathematician and astrologer born at Pavia. As a mathematician he was credited with the discovery of a formula for the solution of certain cubic equations; actually this problem had been solved by Niccolo Tartaglia who had communicated the matter to Cardan, who promised never to divulge the material. Cardan shortly afterwards published it under his own name; and today it is still known as Cardan's formula. In his autobiography, *De Vita Propria*, Cardan reveals his own errors, explaining them as a form of vanity.

Carnot, Sadi Nicolas Léonhard (1796–1832)

French physicist, born in Paris and studied at the École Polytechnique. He did considerable research into heat engines and much of his study led to the discovery of the second law of Thermodynamics. Many disregarded his findings, which Lord Kelvin (*q.v.*) repeated in 1848.

Carrington, Richard Christopher (1826-75)

English astronomer, the son of a Brentford brewer. He worked as an observer at the University of Durham but, finding the way to advancement blocked, he left and set up an observatory of his own at Redhill. For three years he surveyed the heavens within 9 degrees of the North Pole and his findings were published in *Redhill Catalogue of 3,735 Stars*. His principal study was the observation of the motions of sunspots, by which he learnt the elements of the sun's rotation and also found that there was a systematic drift of the photosphere.

Cartel Clock

Originally a French wall clock made of bronze and generally lavishly decorated with figures and plant forms. It became popular in England about the middle of the 18th century.

Cassegrain Telescope

An instrument adapted from the Gregorian telescope by the French scientist Cassegrain in 1672; the main difference is that the small mirror was convex instead of concave. The principle was used with the 4 ft reflector, made in 1870 for the Observatory of Melbourne by the firm of Grubb. The quality of these reflecting instruments relied very largely upon the skill of the opticians and their ability to grind the necessary mirrors.

Cassini

The family name of succeeding Italian astronomers all of whom held the position of director of the Observatory at Paris. *Giovanni Domenico Cassini (1625–1712)* the first of the line was born at Perinaldo near Nice and was educated by the Jesuits at Genoa. After he left he had a varied career. He was nominated as professor of astronomy at Bologna University, wrote a paper on the comet of 1652, he was employed as a hydraulic engineer and inspector of fortifications for Pope Alexander VII. His studies included the rotation periods of Jupiter, Mars and Venus. In 1669 Louis XIV applied for his services for Paris. Some oval curves that he thought to substitute for Kepler's (*q.v.*) ellipses as the paths of planets were named after him – Cassinians. *Jacques Cassini (1677–1756)*, son of Domenico (*q.v.*), was born at the Observatory in Paris and entered the French Academy of Sciences at the age of 17. He measured the arc of the meridian from Dunkirk to Perpignan, supplied tables of the satellites of Saturn in 1716 and wrote *Éléments d'Astronomie* in 1740. *César François Cassini (1714–84)*, son of Jacques (*q.v.*), was also born at the Observatory. He succeeded his father and in 1744 began the construction of a large topographical map of France.

Jacques Dominique Cassini (1748–1845) was also born at the Observatory. He tried to restore and re-equip the establishment but his plans were thwarted by the National Assembly. He resigned and then found himself thrown into prison for seven months. On his release he went to live in isolation at Thury. He completed his father's map which was published by the Academy of Sciences in 1793.

Cassiopeia

A constellation of the northern hemisphere mentioned by Eudoxus (*q.v.*) in the 4th century BC and Aratus in the 3rd century BC. Ptolemy (*q.v.*) catalogued 13 stars in the constellation, Tycho Brahe (*q.v.*) 46, and Hevelius (*q.v.*) 37.

Catenary

The curve assumed by a heavy, uniform section, flexible cord hanging freely from two points. It was investigated by Galileo (*q.v.*), who wrongly determined it to be a parabola; Jungius showed Galileo's mistake but the truth was not found until James Bernoulli published his findings in *Acta Eruditorum*.

Cathetometer

An instrument for measuring lengths and displacements; that is, a graduated vertical pillar along which a telescope can move.

Celestial Equator

The circle in which the plane of the earth's equator meets the celestial sphere; also, the circle on the celestial sphere where it is cut by this plane.

Celestial Globe

It is likely that globes representing the heavenly bodies were made before terrestrial globes. It is possible that Eudoxus of Cnidus (*q.v.*) may have worked with some form of celestial globe. In the Naples Museum there is the oldest existing globe of this type, known as the 'Atlante Farnese', which dates from about 200 BC (there is a half-scale copy in the British Museum). In China, Ho-shing-tien devised a celestial globe in 450. The Arabs were producing examples from about 1050. The globes were usually made so that they could be turned to show what was thought to be the believed rotation of the heavens. With some, the axis could be tilted to match the latitude of the observation point, also there could be a horizontal ring to show which stars would be visible above the horizon at the time of viewing.

Mid–19th century celestial globe by Kirkwood, the paper gores printed with the heavenly bodies, constellations, mathematical and optical instruments

Celestial Sphere

An imaginary sphere with an infinitely large radius which encloses the whole universe so that all celestial bodies appear to be projected on to its surface when the observer is situated at the centre of the sphere.

Celsius, Anders (1701–44)

Swedish astronomer born at Uppsala; he held the chair of astronomy in the University of Uppsala for 14 years. In 1733 at Nuremberg he published papers on his observations, 316 in all, of the aurora borealis. He also did research on the centigrade thermometer. The Celsius scale named after him is the same as Centigrade – 0 degrees is freezing point and 100 degrees the boiling point.

Centrifuge

The name applied to one of a number of machines which, by rotating, can cause the separation of particles in suspension.

Charles, Jacques Alexandre César (1746–1823)

French mathematician and physicist born at Beaugency, Loret. After working as a clerk with a ministry, he turned to scientific pursuits. He was the first to send up a hydrogen filled balloon in August 1783 and in December of that year he and a friend made the first manned ascent with another balloon, rising to a height of 2,000 ft, remaining in the air for two hours and travelling from Paris to Nesle. He worked on the dilation of gases with heat, antedating the research of Gay Lussac.

Charles' Law

From the research of the foregoing, this stated that all gases at constant pressure expand by 1/273 of their volume at 0 degrees Centigrade, for a rise in temperature of 1 degree Centigrade; and the volume of a given amount of gas at constant pressure would be related and proportional to the absolute temperature.

Childe, Henry Langdon (1781–1874)

English inventor who spent much time working on improvements for existing types of magic lantern; in 1818 he introduced a model that had two lamps which made it possible for one picture to be dissolved into the next.

Christie, Samuel Hunter (1784–1865)

The younger son of James Christie (1730–1803), who in 1766 held his first sale at rooms in Pall Mall. Samuel was a mathematician, he studied at Cambridge and became second wrangler. In 1806 he became third mathematical assistant at the Woolwich Military Academy. He invented a torsion balance and researched into electricity and magnetism, publishing papers on the effects of temperature and solar rays on the magnetic needle and the conducting properties of certain metals.

Chromatic Thermometer

A device for measuring temperature by observing the colour of the light radiating from a heated material. As the temperature of a heated mass rises so its colour changes from red to white.

Chromatography

A method of chemical analysis that employs the technique of separating the components

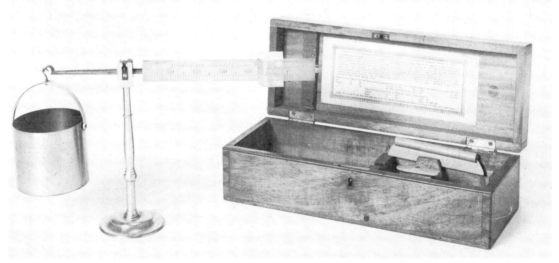

19th century brass chondrometer, the beam signed Young & Son, 5 Bear Street, Leicester Square, *sliding weight engraved* lbs per Bushel Imperial, *steelyard carries a ½ pint capacity bucket*

of a mixture of liquids or gases by selective absorption in a column of powder or on a strip of paper; with the latter the process is called 'paper chromatography'.

Chromosphere

In astronomy, the name given to the shell of luminous gas which surrounds the photosphere of the sun. It has a red tint and during an eclipse it can be seen with the eye at the beginning and the end of the phenomenon; if using a spectroscope observations can be made at other times. The edge of the chromosphere is not regular as huge flame-like masses of highly luminous material shoot out from it; these are termed 'prominences' and demonstrate the high degree of disturbance immediately round the sun.

Chronograph

An instrument for measuring and recording periods of time. It should not be confused with a chronoscope, which measures very short intervals of time. A stopwatch is a chronoscope.

Chronology

The science of computing and adjusting time in order to find out the true historical sequence of past events and their exact date. Chronology differs from history in that it tells of happenings purely with regard to their order in time and ignores their relation to each other. From the earliest recorded periods time has been measured by the revolutions of the sun and moon and other relevant celestial phenomena. The earliest written records have in most cases been lost by deterioration or deliberately destroyed – a Chinese emperor about 220 BC burnt the

relevant books of his time and then buried alive the scholars who knew about them; later a Spanish plunderer destroyed picture records which had been found in a village of Montezuma. The introduction to this book contains further examples of the manipulation of time which occurred until matters were settled to the agreement of all.

Chronometer

A timepiece specially designed to be extremely accurate under any conditions of temperature, movement or atmospheric pressure; especially for use at sea and in connection with navigation. The United Kingdom has led in the development of these reliable instruments. Wallace Nut-

Marine chronometer by Dent, 'Maker to the Queen', Strand and Royal Exchange, London, 19th century

Circumferentor by Thomas Heath, London, 1688

ting, writing in *The Clock Book* published in America in 1924, remarked:

It was natural that Britain, being an island, and therefore necessarily a maritime power, should devote more attention than other nations to chronometers. The reflex stimulus upon clockmaking in general is obvious.

Gemma Frisius (*q.v.*) thought about a shipboard watch in 1530 but it was obvious that the early instruments would have been incapable of withstanding the motions of the sea while remaining sufficiently accurate to assist the navigators. Huygens (*q.v.*), working around 1659, did produce small pendulum clocks that could be mounted in gimbals to counteract the sea's pitching and

rolling. An incentive for development came first from Philip of Spain who in 1598 offered a large reward for the design of such an instrument. In 1714 the British Government offered £10,000, which was followed by £15,000 and then £20,000, for a timepiece that could maintain accuracy on a voyage to the West Indies; for £20,000 the required accuracy was to be within 30 minutes. Celebrated British chronometer makers include Thomas Earnshaw (*q.v.*), John Harrison (*q.v.*), Thomas Mudge (*q.v.*) and Thomas Wright.

Circumferentor

A horizontal compass with diametral projecting arms, each carrying a slit vertical sight, and a dial divided into degrees. The whole is attached to a stand which can be adjusted so that the angle which the line of sight makes with the magnetic north can be observed on the dial. It has been particularly used for surveying in mines.

Clepsydra

A water clock, measuring the passing of time by a controlled flow of water from a cistern or other supply source. The exact date of its development is not certain but it seems that a version was used by the Babylonians, Hindus and Egyptians. The Greeks and Romans apparently regularly employed the device. The simplest form would have probably consisted of a short-necked earthenware globe of a known water capacity, in the bottom of which would have been several small holes through which the water could escape; graduated marks could give some indication of periods of time passing. Another early clepsydra was a copper vessel with a small hole in the bottom; this would be set to float in a bowl of water and when it finally sank a certain period of time would have passed. One use for the clepsydra was to set a limit on speeches in the courts of law. The graduations for measurement with these water clocks were all for twelve periods or hours. The hour, however, was defined as a twelfth part of the time from sunrise to sunset, so the accuracy was impaired by seasonal variations; the devices were also affected by temperature and atmospheric pressure. The most successful model to deal with these difficulties seems to have been developed around 135 BC by Ctesibius of Alexandria (*q.v.*). His instrument apparently relied on a system of water wheels which caused the gradual rise of a small figure which pointed out the hours with a little stick on an index placed on the clepsydra.

Clerke, Agnes Mary (1842–1907)

Irish writer on astronomy, born at Skibbereen, Co Cork, who later settled in London. She was not a practising astronomer but she had a remarkable gift for making the writings of others on the subject readily readable and understandable to the lay public. Her books included *A Popular History of Astronomy During the 19th Century, The System of the Stars, Problems in Astrophysics, The Herschels and Modern Astronomy* and *Modern Cosmogonies*. In 1903 she was elected an honorary member of the Royal Astronomical Society.

Clinical Thermometer

Mercury thermometer designed to measure the temperature of the body; it is graduated to just a few degrees each side of the mean normal temperature. There is a small kink in the tube which causes the thread of mercury to break when the instrument is removed from the body and thus the exact temperature is indicated; it can be reset by shaking.

Clinometer

Used by surveyors and geologists for measuring the dip or angle of inclination of surfaces. The simplest form is just a graduated arc with a plummet. The most useful form of clinometer is one that is combined with a small pocket compass as this will give the direction of the slope as well as its angle.

Clocks

The exact date of the introduction of clocks as known today can only be guessed at. Devices that combine series of wheels and pinions with a weight, pendulum or spring motive force are mentioned in ancient documents, but many of these were produced in monasteries, which complicated matters by using the term 'horologium' too freely and it is difficult to decide whether they are referring to water clocks, sundials or mechanical pieces. European clocks seem to have been introduced during the 13th century, although there is evidence that they may have been invented some centuries earlier and probably displayed as status symbols and curiosities. There is a very limited claim that the balance clock was invented by Pope Silvester II. Other early clocks include one on a former clock tower of Westminster in 1288 with great bells, and having the motto

Late 16th century gilt-metal tower clock, mid-European

German brass hexagonal table calendar clock, movement by Willibald Ittinger; late 17th century

Discite justitiam moniti inscribed upon it; in 1292 there is mention of one in Canterbury Cathedral which cost £30; Dante describes a clock in his *Paradiso* of 1321; there was a celebrated astronomical clock in Strasbourg about 1350; Jacopo de'Dondi in 1344 constructed a timepiece of such apparent excellence that it earned him the title of Del'Orologio; there was the instrument dated 1348 which was in Dover Castle and described in a paper published by Admiral W. H. Smyth (1788–1865) in 1851; the clock itself was included in the Scientific Exhibition of 1876.

Originally clock making was the work of one man with perhaps some apprentices to help him. But from the early 19th century onwards the work became more a matter for teams of specialists in the different parts of the construction. From 1865 parts were being polished by machine and more and more the popular watch for the masses became a matter of mass production.

Cluster

In astronomy, a group of stars which differ from asterisms and constellations in that the members of the group appear to have some physical connection; for example they may exhibit a similar spectrum. One of the better known clusters is Pleiades. Herschel's

English brass monstrance clock, signed James Boyce, London

German brass hexagonal table clock, *18th century.*

English brass chiming skeleton clock, 19th century

(*q.v.*) catalogue of nebulae showed 110 globular clusters. The number of stars in a cluster can be very great indeed; Professor Bailey found some 5,000 stars in a cluster round Centauri, occupying a space similar to that the moon appears to occupy.

Coelostat

An instrument for use in astronomy; it consists of a plane mirror mounted parallel to the earth's axis and rotated about this axis once every two days so that light from a celestial body is reflected on to a second mirror which in its turn reflects the beam into a telescope.

Colatitude

In astronomy and navigation this is the complement of the celestial latitude.

Collimator

An auxiliary telescope used to detect and correct errors in collimation with a larger instrument. Collimation is the adjustment of a telescope in such a manner that the line of sight is set perpendicular to the axis of movement. The collimator is fitted with cross wires and mounted before the transit instrument. When these threads coincide with the axis of the telescope then it will be collimated both horizontally and vertically.

Colorimeter (also Tintometer)

A device for finding the concentration of a solution of a coloured sustance by comparing it with a known and recognized standard.

Comet

A heavenly body of a luminous and seem-ingly nebulous appearance which approaches and recedes from the sun in a highly elliptical orbit. Most consist of three parts: the nucleus and the coma, which together form the head of the comet, and the tail. From earliest observations comets have attracted the superstition of folk-lore – understandably so – as bright headed and trailing a flash of fire behind them, they cruise into our vision from the far reaches of deep space. Comets are very small in mass compared with the planets and they do not appear to exert any gravitational pull on other bodies that they pass. In 1886 a comet passed right through the satellites of Jupiter apparently without disturbing them at all. A strange factor is that the tail of a comet is always directed away from the sun. One theory for this was advanced by the Russian astronomer Professor Brédikhine around the latter part of the 19th century; he felt that there was an electrical action emanating from the sun. As Halley's (*q.v.*) comet once more bears down towards us it is interesting to note that a fiery tailed comet appears on the Bayeux tapestry; experts have worked out the chronology and it seems likely that this was the much travelled Halley's; further, scholars have conjectured that the comet recorded by Chinese astronomers for the years 240 and 87 BC was also Edmund Halley's.

Comet Seeker

A small telescope with a short focal length and a large aperture to give the greatest brilliance.

Compass

Who first devised a mariner's compass is in

Dry card binnacle compass, mid-19th century

some doubt as claims to have been the first have been put forward by the Arabs, the Chinese, the Etruscans, the Greeks, the Italians and the Scandinavians. There are many somewhat vague recordings of sea captains of different countries using a magnetized needle which floated on a small piece of wood in a bowl of water. The evolution of the compass seems to have been very slow. In 1248 Hugo de Bercy notes a change in the construction of compasses, which now have the needle supported on two floats in a glass cup. The earliest reliable description of a pivoted compass is contained in *Epistola de Magnete* (1269) by Petrus Peregrinus de Maricourt. In this the writer describes an improved floating compass with fiducial line, a circle graduated with 90 degrees to each quadrant and movable sights for taking bearings. After this he goes on to talk of a new compass with a needle thrust through a pivoted axis and placed in a box with a transparent cover. When and by whom the familiar compass rose was added is not certain. The Wind-rose is certainly much older than the compass itself, the naming of the eight principal winds goes back to the Temple of the Winds in Athens which was built by Andronicus Cyrrhestes. The earliest known Wind-roses on Mediterranean charts have these winds marked by their initials such as T for Tramontano, G for Grece and S for Scirocco. The north point was indicated on some of the old cards with a broad arrowhead or spear as well as T for Tramontano – gradually these developed into the shape of a *fleur de lis* still common today. The cross in the east continued even on British compasses until around 1700. It is likely that

the naming of the 32 points may have been done by Flemish navigators, but they were familiar even in the time of Chaucer, who wrote in 1391:

Now is thin Orisonte departed in XXiii partiez by thi azymutz, insignificacion of XXiii partiez of the world: al be it so that ship men rikne thilke partiez in XXXii.

The placing of the card at the bottom of the bowl or box was practised by the instrument makers of Nuremberg in the 16th century, and by Stevinus of Bruges around 1600. In 1604 gimbals for suspending the compass to counteract the motion of the ship in rough weather are first mentioned. The term binnacle, originally *bittacle*, is a corruption from the Portuguese *abitacolo*. Evidently improvements were slow, the poor sailors being left very much to fight it out with the elements and part of the time probably having to guess from the results of ignorantly used instruments. In 1820 Peter Barlow reported to the Admiralty that in his opinion half the compasses in the British Navy were no good and should be thrown away. In 1876 the Thomson (Kelvin) compass was introduced.

Compound Microscope
An instrument with several lenses; one of the earliest makers was John Marshall around 1700.

Condensation Pump
One developed to obtain high vacua.

Condenser
A device to concentrate light into a small area or to reduce gases to their liquid or solid form by the abstraction of heat.

Conductor, Electrical

A substance, body or system that will conduct or carry electricity.

Conductor, Thermal

A substance, body or system that will conduct heat or allow heat to flow through it.

Congius

A unit of liquid measurement equal to one Imperial gallon; also, a Roman unit of liquid measurement equal to about 0.7 of an Imperial gallon, 0.84 of a US gallon.

Constellations

Groups of fixed stars generally conceived as representing some mythological figure, the star points being taken as the framework for these figures; thus the partition of the stellar universe and expanse into areas characterized by specific stars can be traced. It is thought that these names originated in the Euphrates valley and were given by the Sumerians and Babylonians. From the earliest times the constellations and the asterisms have been given names connected to phenomena or mythological beliefs. Those early people described the stars as a 'heavenly flock', the sun was the 'old sheep', the seven planets the 'old-sheep stars' and Arcturus, the brightest star in the northern sky, the 'star of the shepherds of the heavenly herds'. The Phoenicians, who were great mariners, appear to have studied the stars for their service to navigation. From the 6th century onwards, legends regarding the constellation subjects were popular with poets and historians. Pherecydes of Athens recorded the legend of Orion, quoting the fact that when Orion sets Scorpio rises; Hellanicus of Mytilene tells the story of the seven Pleiades, the daughters of Atlas; Hecataeus of Miletus gives the legend of the Hydra. In the 5th century BC the Athenian astronomer Euctemon apparently put together a weather calendar in which a number of bodies including Aquarius, Canis Major, Orion, Pegasus and Hydra are mentioned. Ptolemy (*q.v.*) had a catalogue in his *Almagest*. Constellations are featured in Edmund Halley's (*q.v.*) *Catalogus stellarum australium* brought out in 1679. Eleven years later two posthumous works by Johann Hevelius (*q.v.*), the *Firmamentum sobiescianum* and *Prodromus astronomiae* added new constellations. The constant exploration of the universe by enlightened observers with more and more powerful instruments must continue to roll back the shadows.

Converging Lens

A lens that can focus to a point a beam of light passing through it.

Copernicus, Nicolaus (1473–1543)

The great Polish astronomer was born at Thorn in West Prussia, then a part of Poland. After the death of his father he was practically adopted by his uncle, Lucas Watzelrode, later Bishop of Ermland. Nicolaus entered Cracow University and studied astronomy and mathematics under Brudzewo. From there he went to the University of Bologna where he read canon law and attended astronomical lectures by Domenico Maria Novara. He then went to Rome to lecture on mathematics and this was followed by a period studying medicine at Padua. In 1505 he went to Heilsberg, residing at the episcopal palace as his uncle's physician. After his uncle's death he retired to Frauenburg and despite heavy demands on his time from administrative authorities and for a physician's duties he worked hard at the astronomical theories he had begun to develop at Heilsberg.

His great and revolutionary theory was that the sun is the centre of the solar system, with the planets, including the earth, moving round it in fixed orbits and the moon orbiting around the earth. He had been dissatisfied with the findings of Ptolemy (*q.v.*) and ever since his return from Italy had nurtured a firm confidence that the heliocentric theory must be correct. His treatise was virtually complete by 1530 and the news began to spread through *Commentariolus*, a short paper written by Copernicus in that year. Lectures on it were given in Rome by Johann Albrecht Widmanstadt; the approval came through from Pope Clement VII, and Cardinal Schönberg passed to the astronomer a formal instruction for publication of the complete thesis. Slightly 'jumping the gun' George Joachim Rheticus, a close follower of Copernicus, printed in 1540 a partial account of the discovery in *Narratio prima*, at the same time sending to the press at Nuremberg the complete manuscript. This was published in 1543 as *De revolutionibus orbium coelestium* and barely reached Copernicus before he died. It had been dedicated to Paul III, but to a degree the success was robbed by Andreas Osiander, who slipped in an anonymous foreword which misguidedly sought to parry animosity by claiming that the theories presented were purely hypothetical.

Corona

The exterior envelope of the sun composed of extremely hot ionized gases. Until comparatively recently it could only be observed during an eclipse; however, in 1931, Bernard Lyot of the Meudon Observatory in France developed the coronagraph which, if used from a high mountain peak above the haze and dust-laden lower atmosphere, made it possible to see in suitable weather conditions the brighter parts of the corona. The first use of the term corona seems to have been made by Dr Wybord, when observing the eclipse of 1652 which became known as 'Mirk Monday' when it passed over Carrickfergus in Ireland. The first mention in history appears to be in a passage of Plutarch when he refers to an eclipse which was probably the one of AD71. In part he wrote that the obscuration of the moon 'has no time to last and no extensiveness, but some light shows itself round the sun's circumference, which does not allow the darkness to become deep and complete'.

Coronograph

A type of telescope constructed to minimize scattered light and reflected light generated inside the telescope itself. The blackened inside of the tube was further treated to damp down reflection; the main objective lens was mounted in an extra long tube to stop side daylight falling upon it. The solar image formed by the lens was directed on to a silvered disk which was set at an angle to divert sunlight through a window set in the side of the tube. Another single lens, behind the silvered disk, and an arrangement of a diaphragm and small screen, allowed only the light of the corona to pass through to an achromatic lens, which would give an image of the corona that could be observed as free as possible from extraneous and obscuring light and could also be photographed.

Corvus

An ancient constellation described by Ptolemy (*q.v.*). It was sometimes called Hydra et Corvus as it contains part of the body of Hydra. It consists of four principal stars of the second and third magnitude; it is situated below Virgo, between Libra and Crater.

Cosa, Juan de la (*c.*1450–1510)

Spanish cartographer and navigator possibly born at Santona in Calabria. After exploring parts of the western coast of Africa he sailed with Columbus on his famous voyage of discovery in 1492, acting as pilot. He executed two very interesting coloured maps on vellum; one showing the Spanish dominions taken in Africa in 1500, the other setting out the lands discovered by Columbus and his successors.

Cosmic

Relating to the whole universe, infinitely extending into all space and time. The ancient astronomers sometimes used the word 'cosmical' to imply occurring at sunrise. Cosmical physics is a broad inclusive term for cosmical phenomena, terrestrial magnetism, meteorology as related to cosmical causes, the aurora and physical matters concerned with the heavenly bodies.

Cosmogony

The study of the origin and development of the universe or of a certain system in the universe. It is a broad term that includes the myth-making by the early primitive peoples with their ideas of heaven and earth, later to be transformed and formalized by poets, philosophers and priests. Varying tribes and peoples of the past have expressed strange and seemingly widely different ideas, although there is often an undercurrent of similiarity running through them. The Tlingit Indians on the west coast of North America had Yehl, the Raven, who steals a sun, moon and stars from a box to lighten the earth. The Mexicans had five ages, which they called 'suns' for the earth, fire, air, water and the fifth a mystery. In India the *Rig Veda* included a song of praise which began, 'Then there was neither Aught nor Naught.' Another example from India was included in *Manu*.

> ... the self-existent Lord... with a thought, created the waters, and deposited in them a seed which became a golden egg, in which egg he himself is born as Brahmā, the progenitor of the worlds.

Plato thought about a personal Creator, Aristotle proclaimed an uncaused cause and Democritus conceived of a self-created universe.

Cosmography

A study of the representation of the world or the universe; the science which deals with the whole order of nature.

Cottage Clock

Collective term for small Connecticut spring clocks in wooden cases, generally with decorated panels.

Coulomb, Charles Augustin (1736–1806)

French scientist and natural philosopher born at Angoulême and educated in Paris. He investigated the theory of magnetic attraction and repulsion, invented the torsion balance and established the inverse square law of magnetic force.

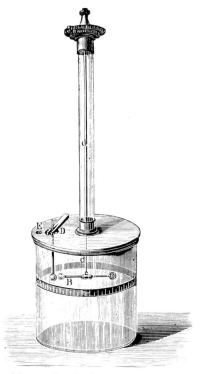

Coulomb's torsion balance

Coulomb

In electricity, the unit of quantity, named after C. A. Coulomb (*q.v.*).

Coulomb's Law

The force of attraction or repulsion between two charged bodies is directly proportional to the product of the charges and inversely proportional to the square of the distance between them. The law can also be applied to magnetic poles.

Coulometer (or Coulombmeter)

An electrolytic cell for measuring the magnitude of an electric charge. This is done by determining the amount of decomposition taking place from the flow of the charge through the cell.

Crepuscular Rays

Those rays that frequently appear when the sun is setting with clouds in its vicinity. The reflection of the light from the clouds and floating dust can often simulate beams of light all coming out from the sun.

Crescent

The name applied to the moon in its first quarter.

Crescent Dial

Self-orienting small pocket directional dial, derived from the universal equinoctial ring dial. The gnomon is crescent shaped and moves on a declination scale. Early 18th-century Augsburg makers Johann Martin and Johann Willebrand produced models.

Crookes, Sir William (1832–1919)

English chemist and physicist born in London, who studied at the Royal College of Chemistry under A. W. von Hofman and then became his assistant. An expert in electricity he invented the radiometer, Crookes' tubes and the spinthariscope – an instrument to pick up traces of radium salts.

Cross Staff

An obsolete device formerly used for taking the altitudes of heavenly bodies, especially the sun.

Crown Glass

A type of glass used in optical instruments; it is less fusible than soda glass as it contains potassium or barium in place of sodium.

Cruciform Dial

A rather complicated multiple universal dial in a cruciform shape, with the arms of the cross forming the gnomons, which are of the same height.

Crystallography

The study of the form, properties and structure of crystals. The first serious step in this science was made by the Danish physician, Nicolaus Steno, who wrote a treatise on his observations of crystals of quartz; this work was published in Florence in 1669 under the title *De solido intra solidum naturaliter contento*. Robert Hooke (*q.v.*) in his *Micrographia*, published in 1665, showed that he had noted the growth behaviour of crystals earlier. The first treatise in full on the subject came in 1723 from M. A. Cappeller with his *Prodromus Crystallographia*. Crystals may be of simple composition – all faces the same – or of combinations of geometrically related faces; research into the varying forms has fascinated scientists and students of geometry. J. F. C. Hessel in 1830 proved that 32 types of symmetry are possible with crystals. A similar conclusion was reached by A. Gadolin, working independently, in 1867.

Ctesibius (*c.*250 BC)

Greek physicist from Alexandria. He was the first to discover the elastic force of air and to harness its strength for motive power. His inventions included the clepsydra, a force pump and a hydraulic organ. The last may have been the greatest of all musical instruments if only for its power; mention of these devices occurs in Italy around the time of Leonardo da Vinci (*q.v.*).

Current Balance

An instrument for the determination of an electric current in absolute electromagnetic units.

Elliot Bros. Dalrymple-Hay's curve ranger; English, c.1900. The sighting telescope is mounted in a trunnion above two spirit levels set at right angles; there is a vertical dial and a plane table.

Cursor

The part of a measuring or calculating instrument which slides or is movable so as to mark a specific position on a graduated or similarly marked scale. With a calculating slide rule the cursor is normally transparent and probably fitted or marked with a cross-hair or fine line. In some devices the cursor is a fine point of light.

Cyclograph

A device for drawing a curve without reference to its centre. It may be in the form of an elastic or pliable strip that can be formed into any curve to fit a set of given points. It is used by, amongst others, draughtsmen and architects when applying some of the classical curves to design. Also called arcograph or curvograph.

Cyclometer

An instrument used by cyclists to show the distance they have travelled. It can be of very simple design. A small stud or projection is attached to one spoke of the front wheel. This engages with a toothed pinion which moves one tooth forward with every revolution of the wheel; this in turn connects to a form of clockwork device which has gears that are related to the circumference of the wheel. The final result is registered in miles travelled on a dial.

Cylinder Dial

The hour-scale is engraved or painted on the outside of a vertical cylinder and the gnomon projects horizontally from the top. It is also known as a shepherd's dial. Some of the simpler ones have been made of wood.

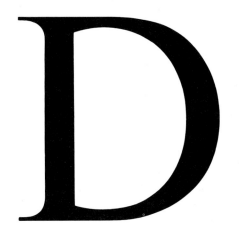

Daguerre, Louis Jacques Mandé (1789–1851)

French painter and physicist born at Cormeilles, Seine-et-Oise. He began his career as an inland revenue officer and then progressed to scene painting and specialized in producing large panoramic spreads, working in conjunction with Pierre Prévost. Later he researched into the possibility of making permanent pictures with the aid of sunlight and eventually he succeeded in producing such pictures on an iodised silver plate. Nicéphore Niepce was also working on the same lines and got in touch with Daguerre, telling him some of his ideas. The two men worked together on their plan for 'heliographic pictures'; sadly Niepce died but Daguerre continued until he was successful. The French Academy of Sciences was so impressed with the importance of the discovery that the government awarded Daguerre the Legion of Honour and annuities were paid to him and Niepce's heir.

Dalen, Nils Gustav (b.1869, active early part of 20th century)

Born at Stenstorp, Sweden, he became a consulting engineer specializing in gas lighting. He developed a device for using this form of power in unattended lighthouses. His other inventions included air compressors, hot-air turbines and milking machines. Unfortunately when working on an experiment he was blinded but was able to continue his mental work.

Dallmeyer, John Henry (1830–83)

German optician born at Loxten, Westphalia, the son of a landowner. When he left school he was apprenticed to an optician in Osnabrück for a time and then in 1851 he went to London and found work with Andrew Ross, the lens and telescope maker. After a difference of opinion with his employer he left the establishment and tried his hand at journalism but gave this up, rejoined Ross and married his second daughter, Hannah. This secured him a legacy when his employer died of one third of an appreciable fortune and the telescope manufacturing department of the business. He became extremely skilled in the making of photographic lenses, object lenses for microscopes and the condensers for optical lanterns. He made successful photo-heliographs for the Harvard College Observatory and for the Wilna Observatory. His second son, Thomas Rudolphus, was the first to make a telephoto lens and wrote an instructional book, *Telephotography,* on its use.

Dalton, John (1766–1844)

English chemist and physicist born at Eaglesfield in Cumberland; his father was a Quaker and a weaver. John Dalton defined an atom as the smallest particle of a substance that can take part in a chemical reaction. He was also an amazingly dedicated observer of the weather; for 57 years he recorded in his diaries some 200,000 observations of the elements. He published *Meteorological Observations and Essays* in 1793. His research also included work on colour-blindness, which he himself had; he could only satisfactorily perceive blue, purple and yellow.

Daniell, John Frederic (1790–1845)

English physicist and chemist born in London. His talents ranged over a wide number of subjects including the Daniell cell, a dew-point hygrometer, a register pyrometer and a water-barometer, which he erected in the hall of the Royal Society. He also developed a gas generated from resin and turpentine which was actually used for a period in New York. His publications included *Meteorological Essays* and *Introduction to the Study of Chemical Philosophy.*

Daniell Cell

Developed by J. F. Daniell (*q.v.*) to provide a steady flow of current, for use with telegraphy. A complicated device, the negative pole stood in dilute sulphuric acid in a porous pot which itself was standing in copper sulphate which also contained the positive pole.

D'Anville, Jean Baptiste Bourguignon (1697–1782)

The celebrated French map maker and geographer who could claim, with some justification, to have started the science of geography. He published 211 maps, the principal collections being *Atlas General, Atlas Antiquus Major, Orbis Romanus* and *Geographie Ancienne abrégée*.

Davy, Sir Humphrey (1778–1829)

English chemist who was born near Penzance in Cornwall and educated at Truro grammar school, where he showed little promise for anything except some quality with his translations of the classics. When his father died, leaving a widow and five children with little to support them, Humphrey set himself to do something for them. He became an apprentice to an apothecary in Penzance and began an intensive course of self instruction on mathematics and chemistry. Reading works by Lavoisier and Nicholson he came up with a fresh theory on heat and light. An early discovery was that nitrous oxide (laughing gas) could be breathed without serious effects; this he proved by breathing some 16 quarts of it for 'near seven minutes'. The gas became popular for obtaining a 'high', Coleridge and Southey being among those who tried it out. He visited the Continent with Faraday (*q.v.*) where they met leading figures in the science world including Ampère, Chevreul, Cuvier and Humboldt. Back in England he devoted much time to investigating 'fire damp' and invented the miners' safety lamp. He graciously took out no patent on this life-saving device and in recognition of this Newcastle coal-owners presented him with a dinner service of silver plate in 1817; this was sold after his death to found the Davy Medal of the Royal Society. He could also write talented verse, so much so that Coleridge commented that had he chosen literature instead of science he could have been 'the finest poet of his age'.

Davy Safety Lamp

An oil lamp for use in mines. It has a cylinder of wire gauze which acts as a chimney, air is admitted at the bottom and when lit the heat of the flame is conducted away by the gauze. This prevents any escape of heat which could ignite the explosive gases outside the gauze. The Royal Institution has a collection of the various models that Davy (*q.v.*) produced during his experimentation.

Day

In astronomy, the period of time it takes the earth on its axis to make one revolution. The days can be distinguished as solar if the revolution is relative to the sun; sidereal if relative to the stars and lunar if relative to the moon.

Declination

In astronomy, the angular distance in degrees of a planet or star from the celestial equator measured north or south along the great circle passing through the celestial poles and the body. Declination can vary year by year; it may also be made to have irregular variations by the action of magnetic storms. Columbus was one of the first to notice that the needle of a compass does not point to the true north.

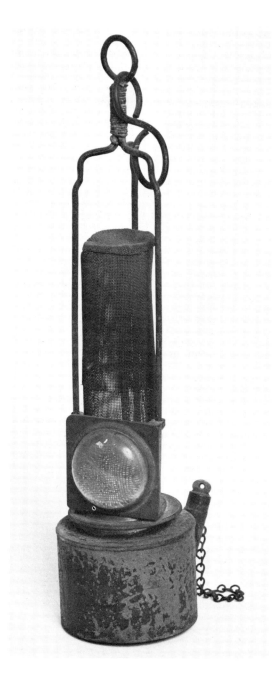

One of the original models for the Davy Safety Lamp

Declination Compass

One that is combined with a telescope (for determining the astronomical meridian), used for finding the declination of the magnetic needle.

Deferent

Ancient astronomers described the deferent as the mean orbit of a planet which carried the epicycle within which the planet revolved. Strictly, it corresponds to the actual orbit of the planet round the sun.

Delambre, Jean Baptiste Joseph (1749–1822)

French astronomer born at Amiens where his studies began, later he continued in Paris and then took a scholarship to the college of Plessis. At first his studies included modern languages, literature and history but then he turned to mathematics and astronomy. In 1790 he was awarded a prize by the Academy of Sciences for his Tables of Uranus. Between 1792 and 1799 he worked on recording the measurement of the arc of the meridian between Dunkirk and Barcelona and published a detailed account of this project in *Base du Système Métrique*, a work that earned him the decennial prize of the Institute in 1910. Amongst his other numerous publications were *Histoire de l'Astronomie Ancienne, Tables du Soleil* and *Astronomie Théorique et Pratique*.

De La Rue, Warren (1815–89)

The son of Thomas De La Rue, the founder of the London stationers; he was born in Guernsey and part of his studies was in Paris after which he joined the family firm but spent most of his leisure time researching into electricity. He was then attracted to astronomy by James Nasmyth (*q.v.*). He made a 13 in reflecting telescope in 1850 and had it erected, first at Canonbury and later at Cranford, Middlesex. With the aid of this instrument he made a number of drawings of heavenly bodies. From this he moved on to work on photographing the heavens and after trials and experiments he made some outstanding photographs of the moon. His inventive and patient mind produced the photo-heliograph which in 1860 he took to Spain to record the total eclipse which occurred on 18 July of that year. In 1873 he retired from active physical observation and gave his instruments to Oxford Observatory. His published works included *Researches on Solar Physics* and *On the Phenomena of the Electric Discharge*.

Densitrometer

A device for measuring the density of tones produced on a photographic plate or film by light, gamma rays or X-rays.

Desiccator

An appliance for drying an over-humid atmosphere; it may be a container filled with a hygroscopic substance such as silica gel.

Dewar, Sir James (1842–1923)

British chemist and physicist born at Kincardine-on-Forth and educated at Dollar Academy and Edinburgh University; he also studied under Kekulé at Ghent. He later became professor of experimental philosophy at Cambridge in 1875 and Fullerian professor at the Royal Institution two years

later. He was the first to liquefy hydrogen and then to solidify it. He erected a device which produced liquified oxygen at the Royal Institution and researched into the characteristics of liquid oxygen and liquid ozone being attracted by a magnet. The properties of gas absorption by charcoal attracted him as he experimented with it at low temperatures. He was president of the British Association in 1902 and was knighted in 1904. His collected papers were published by Lady Dewar in 1927.

Dewar Flask

A double walled glass flask with the space between the walls exhausted to a high vacuum and the inner walls silvered to reduce heat loss or gain by radiation. This seemingly simple device was invented by Sir James Dewar (*q.v.*) and has over the years been used for scientific and domestic purposes.

Diagonal Barometer

An instrument in which the recording arm of the tube is set at an angle of around 45 degrees from the upright member; this allows for a more accurate and sensitive reading as the mercury height will be spread out in the length of the angled tube. It may also be called a Signpost or Yardarm barometer.

Dial

There are numerous types of dial and their classification can often overlap as some of the descriptions refer to the form of the instrument, others to the scientific principle of the dial's design. Entries will be found under the following headings:

Analemmatic; Augsburg; Azimuth; Bloud; Crescent; Cruciform; Cylinder; Equinoctial; Globe; Horizontal; Inclining; Magnetic Azimuth; Magnetic Compass; Mechanical; Pin-Gnomon; Polar; Polyhedral; Regiomontanus; Ring; Rojas; Scaphe; String Gnomon; Vertical; Window.

Gnomonics is the science of dials and their construction. These basically simple devices record the movement of the sun and thus the divisions of the day by a shadow cast on to some form of scale by an upright, an inclined or horizontal member called a gnomon. Probably the earliest mention of a sundial comes in the Old testament in *Isaiah* 38, 8: 'Behold, I will bring again a shadow of the degrees which is gone down in the sun-dial of Ahaz ten degrees backward.' This can be dated at about 700 BC. Early records by Alexander Polyhistor mention the Chaldean astronomer, Berossus, who worked from about 300 to 250 BC and apparently developed a hemicycle or hemisphere, a type of dial that could record the passage of time by a gnomon, in the form of a small bead. The basics of this idea lasted for more than 1,000 years as is evident from the work of Albategnius. In the 18th century a number of dials of this type were found in Italy, including one that may have belonged to Cicero, who wrote in one of his letters that he had sent a dial of the same design to his villa near Tusculum. In the *Almagest* Ptolemy (*q.v.*) discusses the making of dials, horizontal and vertical and facing the different points of the compass. In Athens there was the Tower of the Winds, an octagon on the faces of which were marked the most important winds and relevant dials, the date of these being some-

L. T. Muller brass universal equatorial dial, initialled L. T. M. Augsburg, *early 18th century*

time after the construction of the Tower. The first the Romans would have seen in their city was that erected in 290 BC that had been taken from the Samnites. The first that was actually constructed in Rome was made in 164 BC by order of Marcius Philippus.

The Arabs, in particular, made notable progress with this science; their scholars expanded and diversified the knowledge they had gained from the Greeks. Abu'l Hassan writing around the beginning of the 13th century was probably the first to discuss having equal or equinoctial hours, although like many new ideas it was ignored. In 1531 Sebastian Münster published *Horologiographia*, which contained some basic instructions for the operating and constructing of dials; he also is likely to have invented the first moondial. As the clockmakers of the 18th century produced more and more accurate instruments so the science of gnomonics has lapsed into little more than a mathematical pursuit.

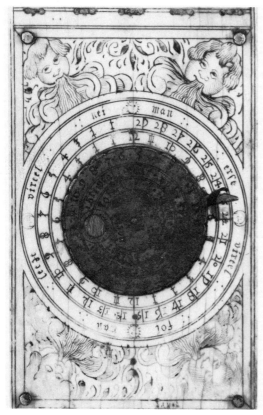

17th century German engraved ivory part-diptych dial bearing maker's mark of two flying birds and signed Joseph Ducher, *dated 1642; the base signed* Joseph Plumen Ducher *and inscribed* 'Angft und Fumer ift aler menfchen Driwfal'

Diamagnetism
Substances such as antimony, copper, gold, lead, silver and zinc that have an extremely low magnetic response are termed diamagnetic as opposed to iron which is attracted to the pole of an ordinary magnet. A. C. Becquered during his research in 1827 noted that even unlikely substances such as wood and some gums could be influenced by magnetic force; in 1845 Faraday (*q.v.*) confirmed this.

Dick, Thomas (1774–1857)
Scottish writer on astronomy, born at Dundee. When he was only nine years old he became inspired by the appearance of a bright meteor which aroused a growing interest in astronomy. Although he had made a start as a weaver, he dropped this and after periods of study at Edinburgh University he became a teacher and later acquired a licence to preach, acting in this capacity as a probationer in the United Presbyterian church. After this he built himself a cottage at Broughty Ferry near Dundee and led a life of writing and scientific study. His books included *Celestial Scenery, The Sidereal Heavens* and *The Practical Astronomer*, which contained some advanced thinking on the possibilities of celestial photography.

Dielectric
A non-conductor of electricity; the term is synonymous with insulator.

Diffraction of Light
A phenomenon caused by interference with a beam of light. If a beam is passed through a small opening or past the edge of an opaque object and then comes on to a screen various patterns and diffusions can be

noted. The rays appear to be bent and to invade the shadow. The condition has been described by a number of scholars including Newton (*q.v.*), Grimaldi and Airy (*q.v.*) although A. J. Fresnel presented the most complete explanation.

Dilatometer
A device for measuring changes in the volume of substances; generally a glass bulb with a long stem which is marked with graduations.

Dioptre
A unit for describing the power of a lens; the reciprocal of the focal length of the lens given in metres.

Dip-Sector
An astronomical reflecting instrument that is in principle very similar to a sextant; it is intended to ascertain the dip of the horizon. In 1803 Wollaston (*q.v.*) spoke of a dip-sector that he had invented, numbers of which were constructed by the firm of Troughton.

Diptych
When applied to scientific instruments, the term implies one made with two folding leaves.

Dircks, Henry (1806–73)
English engineer and author who was born in Liverpool. He studied mechanics and chemistry and is best known for the invention of an optical delusion which was publicly exhibited as 'Pepper's Ghost'. His published works included *Inventions and Inventors* and *Scientific Studies*.

Vertical and horizontal diptych dial by Christopher Schissler (senior), Augsburg, late 16th century

Direct Vision Spectroscope
One that is easily portable and so designed that the eye will look in the direction of the light source when studying the spectrum; there will be no deviation with the centre portion of the spectrum.

Discriminator
An electronic circuit that can convert a frequency or phase modulation into an amplitude modulation.

Dispersion
In optics this term signifies the dividing of

light into rays of different refrangibility; thus when a ray of white light obliquely enters into the surface of a block of glass it will give rise to a divergent system of rays. The simplest way to show dispersion is to refract sunlight either through a glass prism or a special prismatic vessel filled with water or another suitable clear liquid. Sir Isaac Newton (*q.v.*) employed either single prisms or crossed prisms when investigating the phenomenon of refraction or dispersion.

Diurnal Motion

Astronomically, a daily movement completed once every 24 hours – thus the diurnal movement of the earth on its axis in a direction from west towards east.

Diverging Lens

One that can cause a parallel beam of light to spread out when passed through it.

Döbereiner, Johann Wolfgang (1780–1849)

German chemist born near Hof in Bavaria, who, after completing his studies, became professor of chemistry, pharmacy and technology at Jena. He researched particularly into various aspects of combustion. His written work included a treatise on pneumatic chemistry.

Döbereiner Lamp

Invented by J. W. Döbereiner (*q.v.*), this was a device for igniting coal-gas burners and relied on the principle that spongy platinum, when oxygen is present, can ignite hydrogen.

Dollond, John (1706–61)

English optician, the son of a Huguenot

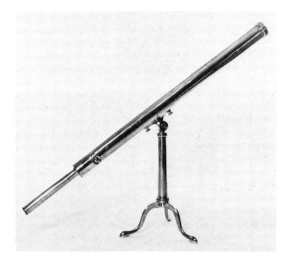

3-inch Dollond brass table telescope, signed Dollond London, *19th century*

refugee; he was born in Spitalfields, London and started work as a silk weaver. In his spare time he studied hard at mathematics, physics, Latin and Greek. In 1752 he gave up his weaving and joined his eldest son, Peter Dollond (1730–1820), who two years previously had set up in business making optical instruments in St Paul's Churchyard. The firm became known as Dollond & Co. His experiments resulted in a number of advances including a means of making achromatic lenses from crown and flint glasses, the production of refraction without colour by the aid of glass and water lenses and later the same result by a combination of different quality glasses. He was also one of the first to construct a refracting telescope though during a patent action the instrument was later discovered to have been privately invented some years earlier – apparently John Dollond's instrument was made as an attempt to prove

Newton's (*q.v.*) theory that such an instrument was possible. This 'achromatic' or 'free from colour effects' telescope is the type normally used for astronomical or terrestrial observation. Dollond's achromatic refractor has not replaced the best Newtonian reflecting telescopes. An account of the life of John Dollond was written by the Rev John Kelly (1750–1809), the Manx scholar, who had married one of Dollond's granddaughters.

Donati, Giovanni Battista (1826–73)

Italian astronomer born at Pisa, who studied at the Observatory in Florence. He became the director of this establishment in 1864. He oversaw the erection of the new observatory in Florence which was completed in 1872. His accomplishments included the discovery of six comets, one of which bears his name, between 1854 and 1864; he subjected the light of a comet to spectrum analysis and thus was able to describe its gaseous composition and made advanced experiments in stellar spectroscopy. He also published *Intorno alle strie degli spettri stellari*, which included ideas for the physical classification of the stars.

Doppler Effect

A phenomenon associated with the apparent change in the frequency of a sound or light vibration caused by the relative movement of an observer and the source of the vibration (also called Doppler shift). The Doppler effect is also used with radar scanning to distinguish between a moving object, informing on its velocity, and a stationary object. The principle was discovered by the Austrian scientist Christian Johann Doppler (1803–53) and has been of considerable value for research and practice in astronomy and physics.

Double Barometer

One which has the mercury column divided into two halves which are fixed to the case side by side and are connected by a tube that is filled with a lighter fluid substance. It was developed from an idea which was worked on by Dr Robert Hooke (*q.v.*).

Draper, John William (1811–82)

American who was born in Liverpool and emigrated to the United States in 1831. He studied at London University and later at the University of Pennsylvania. In 1837 he was elected professor of chemistry in the University of the City of New York. His most important research lay with photochemistry; he made advances with Daguerre's (*q.v.*) process. Other work included the discovery that on reaching about 525 degrees Centigrade all substances glow a dull red, the temperature reading being known as the Draper point. His publications included *Textbook on Chemistry* and *Scientific Memoirs*.

Drawing Pens

The draughtsman's pen which is expressly for ruling lines is made in the form of two flat pointed blades that can be compressed together to make thin lines by adjusting a small screw wheel. The pens are filled with ink by using a small dropper or fine brush. Sometimes one of the blades is hinged so that the pen can be opened and thus cleaned more easily. An ink pen for a compass is made in the same way.

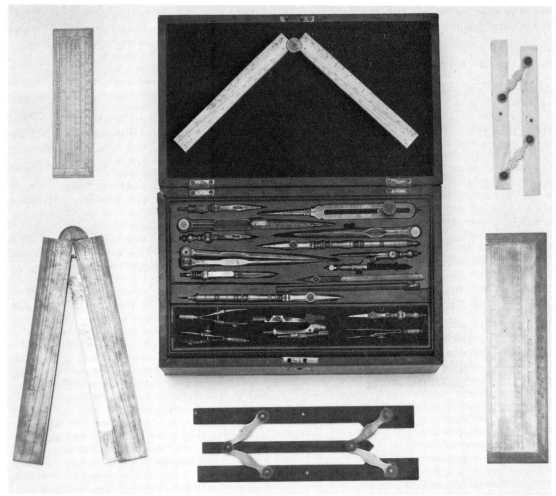

Mid-19th century naval architect's drawing instrument set contained in a fitted mahogany case with engraved plaque R. Bishop, Northfleet Dockyard 1852. *Set comprises brass and steel compasses, triple parallel rule in ebony, scales, ivory sector, dividers, proportional compass, pens and ivory rule, signed* Fraser, New Bond Street, London.

Dutch Striking

A form of striking in which the hours are repeated at the half-hour by the sounding of a bell with a different tone.

Dynamo

A machine for converting mechanical energy into electrical energy. One of the outstanding applications of electrical and magnetic science, the principle was discovered by Michael Faraday (*q.v.*) in 1831.

Dynamometer

An instrument for measuring force in terms

of brake horsepower. It may be called a brake or absorption dynamometer.

Dyson, Sir Frank Watson (b.1868, active until 1933)

English astronomer born at Ashby-de-la-Zouch and educated at Bradford grammar school and Trinity College, Cambridge. His appointments included chief assistant of the Royal Observatory Greenwich, secretary to the Royal Astronomical Society, Astronomer Royal for Scotland and finally Astronomer Royal at Greenwich in 1910.

E

Earnshaw, Thomas (1774–1829)

English clockmaker from Ashton-under-Lyme; he was possibly apprenticed to a country watchmaker and later when he set out on his own was hampered in his research into chronometers by lack of funds. To solve this, he made a living in London making clock parts for such makers as John

Brockbank and Thomas Wright. His inventive skill led to the introduction of simplified chronometers that contain the basics of those made today.

Echo Sounder

An instrument for finding the depth of water by measuring the time taken for a pulse of high frequency sound to travel to the sea bed or a submerged object, and for the echo to return to the measuring instrument. It is also called Asdic or Sonar.

Eclipse

The total or partial obscuring of one heavenly body by the shadow of another. There are three types of the eclipse – those of the sun, the moon and the satellites of other planets.

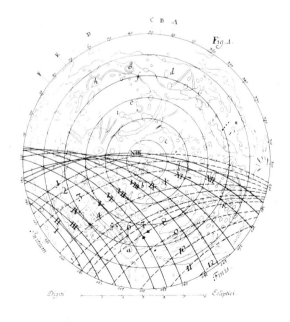

The eclipse of the moon on 19 November, 1686, recorded in Acta Eruditorum, *Leipzig*

Ecliptic

In astronomy, the great circle on the celestial sphere which represents the apparent annual path of the sun relative to the stars. The path lies through the middle of the constellations known as the Signs of the Zodiac. The obliquity of the ecliptic is the angle which its plane makes with that of the Equator.

Edison, Thomas Alva (1847–1931)

American inventor born at Milan, Ohio of Dutch and Scottish descent. Shortly before he was eight the family moved to Port Huron on Lake Michigan; here the young boy went to school, but after three months the teachers declared that he was exceptionally unintelligent and always at the bottom of his class, and from hereon his mother taught him. At the age of 12 he became a newsboy on the trains to Detroit; whilst so occupied he managed to print a small newssheet and also to do some chemistry experiments to assuage a rising interest in invention. When he was 15 he trained himself to become an extremely efficient telegraph operator. He travelled through many cities, eventually reaching Boston in 1866; here he set up a small workshop and constructed what was probably his first invention, a vote-recording machine which he patented. In 1869 he moved to New York and after a time set up another work place, at Newark, New Jersey. He produced an improved printing telegraph for stock quotations for the Gold and Stock Company, for which he was paid $40,000; the young genius was so amazed that according to record he carried the cheque about with him for days, not being quite sure what to do with it. His

incredible catalogue of inventive work included a microtasimeter for detecting small changes in temperature, the phonograph, the megaphone, the incandescent light, the kinetoscope – a carbon telephone transmitter, the thermionic valve used with radiotelegraphy and radiotelephony, a

Edison's electric lamp, 1880

system for wireless telegraphy between moving trains, alkaline storage batteries and an advanced dynamo. For over 50 years ideas poured from him, and he amassed about 1,000 patents. Henry Ford said of his countryman:

An inventor frequently wastes his time and his money trying to extend his invention to uses for which it is not at all suitable. Edison has never done this. He rides no hobbies. He views each problem that comes up as a thing of itself, to be solved in exactly the right way ... His knowledge is so nearly universal that he cannot be classed as an electrician or a chemist – in fact Mr Edison cannot be classified ... The more I have seen of him the greater he has appeared to me – both as a servant of humanity and as a man.

Eidograph
An instrument that can be used instead of a pantograph for copying drawings, or for enlarging or reducing them.

Electrodynamometer
A device for measuring voltage, power or current in direct current or alternating current circuits.

Electromagnet
A temporary magnet. It is simply made by winding a coil of wire round a piece of soft iron. When current is passed through the coil the iron becomes a magnet, when the current is switched off the magnetic property is reduced.

Electrometer
An instrument for measuring differences in voltage. Whilst in use it does not draw current from the circuit.

Electron
The basic corpuscle or 'atom' of negative electricity, the name was suggested by Dr Johnstone Stoney in 1891. Professor J. J. Thomson (q.v.) proved in 1897 that cathode rays consisted of these 'corpuscles', as he termed them. Electrons can be given off from metal surfaces when illuminated by ultra-violet light, from incandescent metals at extreme temperatures and from certain oxides.

Electron Lens
An arrangement of electrodes or magnets which can focus a beam of electrons in a manner similar to an optical lens. The idea is used in electron microscopes.

Electron Micrograph
The name for a photograph obtained with an electron microscope.

Electron Microscope
A high power instrument, similar in purpose to the normal light microscope except that it employs a beam of electrons instead of a beam of light. The resolution is considerably better.

Electroscope
A sensitive device for detecting the presence of an electric charge. The basic construction is a rod holding two pieces of gold foil that separate when current is applied.

Electrostatics
The study of electricity at rest, static, as opposed to the study of moving electric currents.

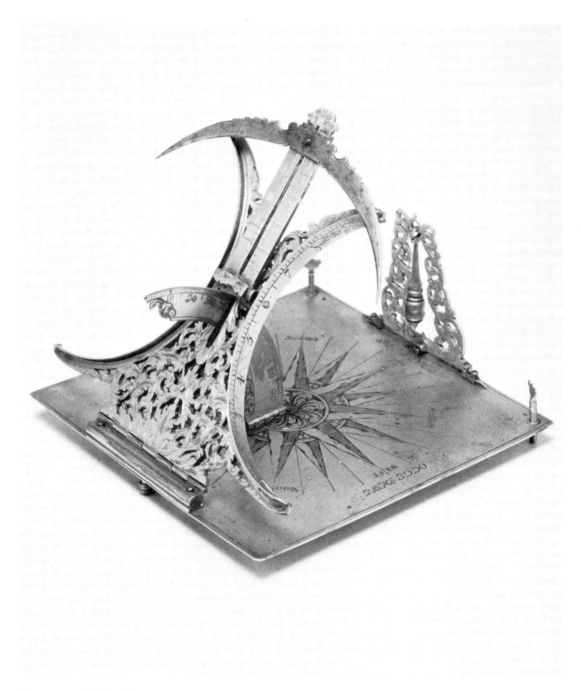

Equatorial dial, silvered and gilt, by John Willebrand, c.1700

Ellipse

A regular oval; it can be produced when a cone or cylinder is cut through at an angle.

Elliptical Compass

Not a drawing compass in the strict sense. It is a device for accurately producing an ellipse. It works by putting a pen or pencil through a hole in a sliding rod that can be so set that it will trace an ellipse to any given width and length.

Epicycle

In ancient astronomy, a small circle which was supposed to move on the circumference of a larger one. It was employed to represent geometrically the motions of such heavenly bodies as Mars, Jupiter and Saturn. Ptolemaic (q.v.) theory accepted the movements as real, but as astronomers understood the Copernican (q.v.) theory, the former was rendered void.

Epidiascope

An optical projector for throwing a magnified image of either an opaque three dimensional object, a transparency or a drawing on to a screen. Some models allow a hand with a crayon or pencil to work on the surface being projected.

Equator

In geography, the great circle round the earth which is equidistant from both the north and south poles, dividing the northern hemisphere from the southern hemisphere. This Equator is termed the terrestrial as opposed to the celestial equator which in astronomy is the name given to the great circle in which the plane of the

Brass equatorial dial, signed Baradelle Paris, *18th century*

terrestrial Equator intersects the celestial sphere; the celestial equator is consequently equidistant from the celestial poles. The so-called 'magnetic equator' is an imaginary line circling the earth along which the vertical component of the earth's magnetic force is zero; it lies very close to the terrestrial Equator.

Equatorial Telescope

An astronomical instrument mounted on an axis perpendicular to the plane of the Equator. It is moved by machinery to keep the stars on which it has been set always exactly in the field of view during their apparent movement across the heavens. Photographs taken with the aid of such an instrument show the stars as sharply defined points of light, whilst asteroids or comets are shown as short white streaks.

Equinoctial Dial

One in which the hour-scale is set parallel to the plane of the Equator. As the scale is circular it can very easily be divided with accuracy.

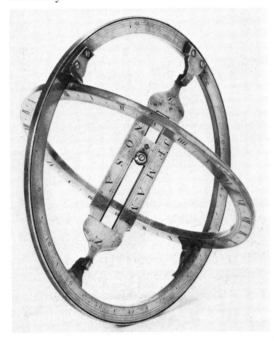

18th century brass universal equinoctial ring dial. Bridge engraved with the signs of the zodiac; back of the meridian ring has a scale for determining the Sun's meridian altitude.

Equinox

The moment when the sun apparently crosses the Equator. This happens twice in the year, on 21 March and 22 September, when the days and nights are of equal duration all over the world. At the vernal (March) equinox the sun passes from south to north and the movement is reversed at the autumnal equinox. Popular belief has it that during the two equinoxes great gales can be expected, but long term observation has proved that there is no foundation for this.

Eratosthenes of Alexandria (c.276–c.194 BC)

Born at Cyrene, he studied under Callimachus at Alexandria and also at Athens. At the summons of Ptolemy III he returned to Alexandria where he was appointed chief librarian. One of the great scholars of antiquity, he wrote on a number of subjects but sadly much of this has been lost, although fragments of his astronomical poems remain. Amongst his accomplishments were the measuring of the earth, the circumference of which he calculated as 252,000 stadia, a system of scientific chronology having as its base for calculation the conquest of Troy and a theory by means of which prime numbers might be discovered.

Erecting Prism

One with a right angle, employed with optical instruments to bring an upside down image upright.

Escapement

A device used with timepieces to give periodic impulses from the spring or weight to the balance or pendulum; it usually consists of an escape wheel and anchor. Invented around 1670 by William Clement, this revolutionized the accuracy of timekeeping.

Eudiometer

A graduated glass tube that is used to study and measure the volume changes during chemical reactions between gases.

Eudoxus of Cnidus (c.406–c.355 BC)

Greek astronomer who studied for a period under Plato and later with priests of Heliopolis in Egypt; he then taught physics in Cyzicus and the Propontis, later going back again to Athens with some of his pupils where he opened a school which was to rival that of Plato. Vitruvius credits Eudoxus with the invention of the sundial, and Strabo (q.v.) stated that he found that the solar year was 6 hours longer than 365 days.

Large brass universal equinoctial dial for the Russian market, signed Priestley, *late 18th century*

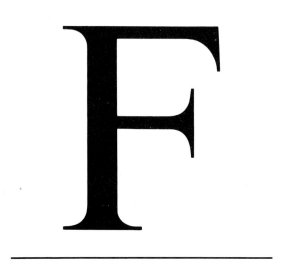

Fahrenheit, Gabriel Daniel (1686–1736)

German physicist born at Danzig, he spent most of his life in England and Holland researching into physics and making meteorological instruments. One of his inventions was for an improved form of hygrometer; he also did much to improve the construction of thermometers and he introduced the thermometric scale known by his name.

Fahrenheit Scale

A temperature scale in which the melting point of ice, or freezing point of water is set at 32 degrees and the boiling point of water at 212 degrees.

Fan or Fly

A rapidly revolving vane in the striking train of a clock, which acts as a governor controlling the rate of strike.

Faraday, Michael (1791–1867)

English chemist and physicist born at Newington, Surrey, the son of a blacksmith. He was apprenticed to a bookbinder but spent all his spare time studying and experimenting as best he could with limited means and equipment. He attended some of the lectures given by Sir Humphrey Davy (*q.v.*) who, impressed with his sincerity, made him an assistant in the laboratory of the Royal Institution. When Davy travelled through Europe he took the young Faraday with him. In 1825 he was appointed as director of the laboratory and in 1833 Fullerian professor of chemistry in the Institution for life, with the added comfort of not having to deliver lectures. Faraday's early chemical work followed Davy's lines and he particularly researched into chlorine and the liquefying of gases and produced new kinds of glass for optical purposes. But his main contributions were his electrical discoveries. His interest in this field started in 1812 when he made his first battery from seven half-pennies, seven pieces of zinc and six pieces of paper moistened with salt water. In 1821 he began his work on electromagnetism and made his first important discovery, which was the production of the continuous rotation of magnets and of wires conducting the electric current round each other. In 1831 he made his second triumph with the understanding of the induction of electric currents and later in that year he constructed a dynamo which consisted of a disk of copper revolving between the poles of a horseshoe magnet; two rubbing contacts closed the circuit and a galvanometer registered the continuous current. Faraday was totally devoted to his studies and sincere in the presentation of his researches. Dr Bence Jones commented on him:

His standard of duty was supernatural. It was not founded on any intuitive ideas of right and wrong, nor was it fashioned upon any outward experiences of time and place, but it was formed entirely on what he held to be the revelation of the will of God in the written word, and throughout all his life his faith led him to act up to the very letter of it.

Faraday's published works included *Experimental Researches in Electricity, Chemical Manipulation, being Instructions to Students in Chemistry, Lectures on the Chemical History of a Candle* and *On the Various Forces in Nature*, in which he speculated on the possibility of dispensing entirely with the atomic theory, in the place of which he would propose a centre of force theory.

Fathometer
An instrument for measuring the depth of water which uses the echo sounding principle.

Ferrel's Law
This states that everything moving on the earth's surface is subject to a deflecting force caused by the rotation of the earth. The deflection is to the right in the northern hemisphere and to the left in the southern hemisphere. The law is named after the American meteorologist, William Ferrel (1817–91).

Ferrotype
A photographic process for developing negatives by using a solution of protosulphate of iron and gum arabic. It was developed by Robert Hunt in 1844 and has also been called Energiatype.

Field
The area or region in which an electrically charged or magnetized body can exert its influence.

Field Coil
A coil of wire which can magnetize an electromagnet, as with a dynamo.

Field Lens
That one in the eyepiece system of an optical instrument which is set farthest from the observing eye.

Field Magnet
That which provides a magnetic field with an electrical motor.

Finder Telescope
One that is of low power that can be fixed parallel to a large telescope to facilitate the location of the object to be observed and the placing of it in the field of vision of the more powerful instrument. It is normally used only in conjunction with astronomical observations.

Fitz, Henry (1808–63)
Born in Newburyport, Massachusetts, the son of a hatter. When Henry was 11 the family moved to New York where he was apprenticed to a printer. From this he moved to locksmithing with William Day of New York. Between 1830 and 1839 he travelled the building sites in New Orleans,

Philadelphia and Baltimore selling locks in an endeavour to make enough money to start a locksmith's business of his own. He never rode if he could walk, neither drank nor smoked, ate little meat and lived chiefly on water and Graham bread (made from wholewheat flour and named after Sylvester Graham, an American vegetarian who urged dietary reform). Quiet evenings were spent in reading and hobbies, the chief of which was astronomy. His diaries and letters show that when financially able he purchased telescopes and lenses. His first instrument was made in 1838, a reflector telescope with which he delighted in showing his friends the stars and planets. The Rev Clapp of New Orleans told one meeting that he was 'the young locksmith who knew more about the heavenly bodies than anyone else in the United States'. But money was hard to come by to forward his experiments. In 1839 he was working as a speculum maker with Wolcott and others. Later that year he was working with a camera invented by Wolcott and produced a portrait which is believed to be the first ever made. For a time he had a studio in Baltimore and made a few dollars to help the telescope business to progress; his first refractors were made there but later he was to describe them as crude affairs. The stimulus he needed for harder work came when he married Julia Ann Wells of Long Island. Fired by encouragement they moved to New York where they were to settle for the rest of his life. In 1845 he constructed a 6 in refracting telescope for showing at the Fair of the American Institute which was held annually in New York. It had an achromatic objective and a novel

tripod and it won for him the Gold Medal, the highest award at the Fair. Lewis M. Rutherford, a trustee of Columbia College ordered a 4 in refractor for his private observatory. Although he still made cameras, more and more Fitz devoted himself to those instruments that could help others to see a way to the stars. His search for perfection turned him towards the right glass. For the 1845 model he had used a Boston-made flint with French plate glass, but this proved too veiny for large lenses so he imported crown and flint glasses. In 1856 he worked on larger telescopes – a 12½ in refractor and later a 16 in telescope for Mr Vance Duzee of Buffalo; a 13 in for the Allegheny Association at Pittsburgh and a 12 in for Vassar and the University of Michigan. Finance built up and the family was able to stop renting and to build a house in 11th Street. Tragedy struck shortly after they moved, however, when a heavy chandelier fell on Henry Fitz and mortally wounded him leaving his son Harry to carry on the business. Several of Henry's instruments, though made over a century ago, are still functioning.

Fixed Stars
Specifically, refers to those stars in the Ptolemaic (*q.v.*) system which, by imaginative thinking and romance, were assumed to be attached to some vast outer crystal sphere which explained to the ancients their apparent lack of movement. The term is also applied to some distant star which appears over a lengthy period of time not to be moving. The old idea that the stars were absolutely fixed has long since been shown to be inaccurate, although the 'fixed stars'

still survive in many people's astronomical vocabulary. Dedicated observation over a number of years has shown slight changes of position even with the most seemingly 'motionless' of these bodies – evidence which points to these far-off bodies having 'proper motions' in space.

Flamsteed, John (1646–1719)

English astronomer, born at Denby near Derby, the only son of Stephen Flamsteed, a maltster. John attended the free school in Derby but left fairly soon with a physical handicap. Whilst travelling to Ireland he was 'stroked' by a certain Valentine Greatrakes but he was left with a long resting period during which he studied the science of astronomy; he read every book on the subject that he could lay his hands on and turned to making some simple instruments, including some for measuring. The various papers which he wrote gradually brought him acclaim. In 1670 he entered Jesus College, Cambridge and took his degree four years later. In 1675 he was invited to London by Sir Jonas Moore who proposed to set him up in a private observatory at Chelsea. But Charles II had other desires for the talented Flamsteed; the monarch wished the tables of the heavenly bodies to be corrected and the positions of the fixed stars to be righted 'for the use of his seamen'. Flamsteed received a Royal Warrant on 4 March 1675. In his own way, he was a perfectionist and sought to keep all the results he was obtaining until they could be published in a proper manner, in this he crossed Newton (*q.v.*) who depended on punctual information from the first Astronomer Royal for work on his lunar theory.

Flamsteed's publications included the fine three volume work *Historia coelestis Britannica* and the *British Catalogue* of nearly 3,000 stars.

Fleming, Sir John Ambrose (1849–1945)

English engineer born at Lancaster, who, after a brilliant academic career, was appointed university demonstrator at St John's, Cambridge. Later he moved to Nottingham University and then became electrical engineer to the Edison Electric Lighting Company in 1881. He was the first to develop a thermionic valve. His publications included *Magnets and Electricity Currents*, *Radio-telegraphy and Radio-telephony*, *The Thermionic Valve in Radio-Telegraphy* and *The Interaction of Scientific Research and Electrical Engineering*.

Flint Glass

A type of high quality glass which contains lead silicate and is suitable for optical instruments.

Fluorescence

A characteristic of some substances to emit light of a certain wavelength after being exposed to light of another wavelength but ceasing to emit when the source of light is cut off. This phenomenon was first described by Sir David Brewster (*q.v.*) when lecturing to the Royal Society of Edinburgh in 1833; the name he gave to it at that time was 'internal dispersion'. A few years later Sir John Herschel (*q.v.*) independently found that if a solution of quinine sulphate which, under transmitted light, appeared transparent and colourless like water, was illuminated by a ray of daylight a strange blue colour could be noted in the top layers of the solution. Fluorescent substances include fluor-spar, chlorophyll, uranium glass, paraffin oil, tincture of turmeric and magnesium platinocyanide.

Fluoroscope

A fluorescent screen which enables the direct observation of X-ray images.

Fluxmeter

An instrument for measuring magnetic flux; basically a moving coil galvanometer.

Focal Length

The distance from the surface of a lens or mirror to the focal point.

Focus

A point to which something converges or diverges. Also, the sharpness and clarity with which an optical system with a microscope, telescope, camera or projector can render an image.

Foot-Candle

Unit of illumination equal to one lumen per square foot.

Foot-Pound

Unit of energy or work equal to the work done by a force of one pound moving through a distance of one foot.

Form Watch

One that is made to a design or in a manner that is out of character for the period in which it was made. Some makers have incorporated their movements into such unlikely mounts as skulls, musical instruments and baskets of flowers.

Fortin Barometer

With this instrument before the reading by difference is taken, the level of the cistern is set to a fixed pointer, thus capacity error is eliminated.

Foucault, Jean Bernard Léon (1819–68)

French physicist born in Paris, the son of a publisher. He was primarily educated at home, until he determined to make a career in science. For three years he was experimental assistant to Alfred Donné (1801–78), with a course on microscopic anatomy. One of his first pieces of inventive research was an improvement of the Daguerre photographic process. Later he worked with A. H. L. Fizeau on investigations on the intensity of the light of the sun compared with that of the carbon arc and of lime in the flame of the oxyhydrogen blowpipe. In 1844 he developed an apparatus for using electric light in optical experiments and microscopes; the following year he demonstrated the earth's rotary motion by means of a pendulum. Between 1851 and 1852 he invented the gyroscope and in 1857 he invented the polarizer which bears his name. In 1862 using Wheatstone's (*q.v.*) revolving mirror he determined the absolute velocity of light to be 298,000 km a second, some 10,000 km less than the result obtained by earlier experimenters. From 1845 he edited the scientific portion of the *Journal des Débats*; his principal scientific papers were published in the *Comptes Rendus 1847–1869*.

Fournier D'Albe, Edmund Edward (1868–1933)

English physicist born in London, he became lecturer at Birmingham and later at Lahore. His inventions included in 1912 the optophone, a device to enable the blind to read, a system of wireless telewriting and telephotography. On 24 May 1925 he transmitted the first wireless picture to be broadcast from London. His written works included *The Electron Theory, The Moon Element, Quo Vadimus?* and *Hephaestus, or the Soul of the Machine.*

Franklin, Benjamin (1706–90)

American statesman, author and scientist, he was born in Boston, Massachusetts, the fifteenth child in a family of 17. At the age of eight he was sent to attend the Boston grammar school, but when only ten he left to assist his father, who worked as a tallow chandler and soapboiler. Three years later he was working with his half-brother who was a printer. In 1725 he sailed to England where he worked for a year and a half in a printer's office. On his return to America he set himself up as a printer and in 1729 bought the *Pennsylvania Gazette*, with which he was so successful that he expanded by bringing out *Poor Richard's Almanac*.

His scientific investigations included the establishment of the identity of lightning with electricity, this he did by means of his celebrated experiment with a child's kite, the placing of lightning conductors on large buildings, the discovery of the Gulf Stream and the course of the storms that crossed the North American continent, proposals for watertight compartments for ships, floating anchors to steady a ship in a storm, crockery dishes that would not upset during rough sea passages and in 1784 he was a member of the committee which investigated Mesmer, the report of which is of lasting scientific value. Franklin was in favour of vegetarianism and a simple diet, temperance and good ventilation and, for himself, he invented a pair of bifocal glasses.

Called by many the Patriarch of the United States, he can perhaps best be summed up with the words of Bernard Fay:

In the eighteenth century no one cared a great deal if Franklin had invented scientific or philosophical theories, a religion, or political principles. It was he, himself, they loved; it was the magnificent role he created for himself...that attracted people. No one had ever been able before to play the bourgeois, but he had known how.... He had known how to discard absolute beauty, absolute truth, absolute good, as well as evil, voluptuousness, frivolity and elegance: he had only kept hold of the practical and had given it a human, attractive and picturesque form.

Franklin's written work was not collected during his lifetime nor did he make any effort to have it published.

Franklin Clock

A type of shelf clock with a wooden movement made by Silas Hoadley about 1825. With another clockmaker, Seth Thomas, Hoadley had bought out Eli Terry, the famous clockmaker who had been largely responsible for factory production of clocks in America. Terry's fortune was made when, owing to the shortage of metal because of the blockade by Napoleon, he received an order for 4,000 wooden clock movements which he was able to fulfil.

Fraunhofer, Joseph Von (1787–1826)

German optician and physicist, born at Straubing in Bavaria, the son of a glazier. In 1799 Josef was apprenticed to Weichselberger, a glass-polisher and maker of mirrors. Shortly after this he had a strange stroke of fortune; the house in which he was lodging collapsed and he nearly lost his life. When he was being dug out the Elector of Bavaria, Maximilian Joseph, was watching and presented the astonished youth with 18 ducats. This sum enabled him to purchase his freedom from the apprenticeship and set up his own glass workshop. In 1806 he got the position of optician in the mathematical institute which had been founded in 1804 in Munich by Joseph von Utzschneider. Whilst there he studied with the leading optician Pierre Louis Guinand and learnt how to make flint and crown glass. His inventions and developments included a device for polishing mathematically exact spherical surfaces, a stage micrometer, a type of heliometer, long focus achromatic lenses for microscopes and the large reflecting telescope at Dorpat. Amongst his discoveries were the dark lines in the sun's spectrum which bear his name. Much of his research was published in the *Denkschriften der Münchener Akademie* for 1814–15.

Frisius, Gemma (1508–55)

The astronomer whose workshop was the inspiration to many including Gaulterus Arsenius of Louvain, a talented maker of astrolabes. Gemma himself produced a number of excellent instruments including a small portable equatorial armillary suitable for suspension by hand. He also revived a method of projection which was recorded by the Andalusian astronomer al-Zarqāllu (*c.*1029–87), which would make a single instrument practical for all latitudes. In 1530 he wrote a treatise on both astronomy and cosmogony, in which he dealt with meridians and the establishing of longitude and latitude, and scales for measuring distances. The volume must have been popular as it was later translated into French by Claude de Bossière. Four years later he made an 'astronomical ring', largely from the design of someone else, that he claimed would do all that astrolabes, cylinders and quadrants had done.

Galaxies

Star systems held together by gravitational attraction either as a regular galaxy which may be in a symmetrical shape, elliptical or spiral, or as an irregular galaxy with asymmetric shape. There are billions of galaxies existing outside our stellar system.

Silver and gilt bronze armillary sphere by Jakob Mannlich, Troppau and Augsburg, c. 1645

I

Mid-17th century brass Maghribi astrolabe

Two astrolabes forged c. 1978. The standard of craftsmanship is high, but a similarity of engraved letters and figures on numerous examples from the forger's workshop pointed to the deception.

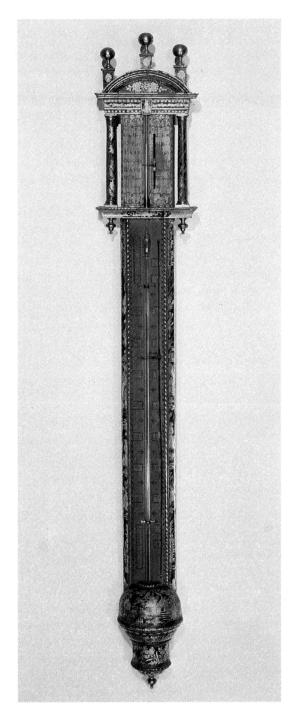

Movement of late George II longcase clock
with solar and lunar disc by Nickals, Wells

Oak stick barometer in the style of John
Patrick, early 19th century

South German Madonna automaton clock by Binedickh Miller, 17th century

Thomas Tompion made this year-going spring clock for William III, c.1695-1700

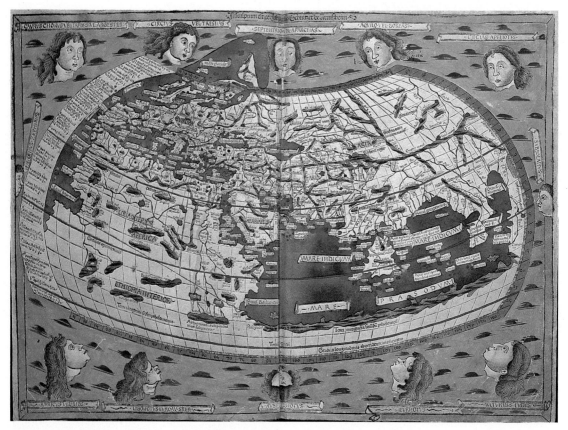

Ptolemy: Cosmographia, *from the second German edition, hand-coloured on a woodcut. Ulm, J. Reger, 1486*

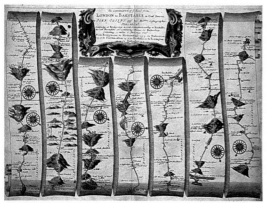

Road map by John Ogilby, c. 1675, from his 'Britannia', the earliest road route maps of England

Rare German gilt-metal chalice dial signed Marcus Purman Monachi Faciebat 1608. *Purman was active 1583-1616; in the early part of this period he worked under Ulrich Schniep, instrument maker to the Dukes of Bavaria. The interior of the chalice bowl is calibrated for 48° latitude, with scales for the hours IV-XII-VIII and the sun's altitude. There is a central gnomon, and the dial is for use when the chalice is filled with water. The inscription round the inner ring reads:* WAN ICH PIN : EIN GESCHENCKT OBEN VOL * SO ZAIG ICH DIE STUND GAR WOL * PIN ICH ABBER LEHR * SO DUE ICH ES NIT MERR**** *One of the circular bosses on the base swivels to reveal a magnetic compass.*

Top left: German gilt-metal and silver inclining dial, signed J. E. Essling Fecit Berolin 1716
Top right: German ivory diptych dial, maker's mark H. D. with a snake between, c. 1580
Bottom left: French ivory diptych, possible maker's mark of P I, 1565
Bottom right: Gilt-metal and silver compendium signed Elias Allen fecit, *first half of 17th century. Allen is recorded as having been working in London from c. 1600. The instrument still retains the original mica glass.*

Polychromed wood compendium dial and nocturnal with silver fittings inscribed Horarium Generale A Solle Luna et Stellis At Solvebat Hilarius 1556. Face 1: *Nocturnal with incised concentric calendar and twice XII hour scales, toothed silver rundle calibrated V-XII-VII and folding silver index.* Face 2: *Vertical pin gnomon twice XII solar dial with concentric zodiac ring.* Face 3: *Horizontal solar dial with inlaid ivory chapter scale calibrated IIII-XII-VIII, inset magnetic compass, silver gnomon folding over central lunar volvelle dial.* Face 4: *Silver plate engraved with latitudes of various towns.*

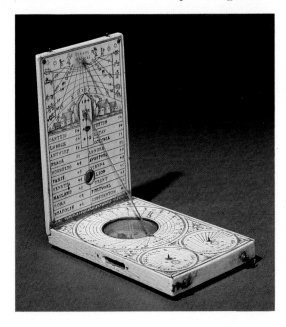

17th century ivory German diptych dial with red and green stained decoration, string gnomon, pin gnomon dials, table of latitudes and compass, bearing the fleur de lys *trade mark of Leonhardt Mire of Nuremberg*

German compound monocular microscope, mid-18th century

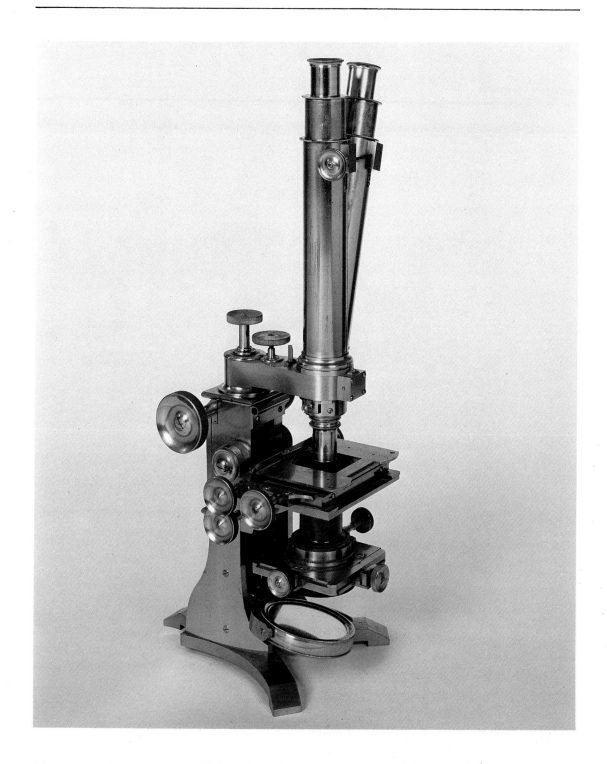

19th century brass compound binocular microscope by Baker, High Holborn, London

Early 19th century single and compound chest microscope signed Dollond, London, *with rack and pinion stage adjustment*

Unusual 18th century brass pedometer, with silvered dial signed Nairne and Blunt, London

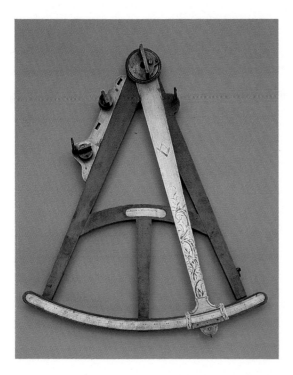

*18th century ebony and brass mounted
quadrant by John Goater*

*18th century pasteboard and brass mounted four-draw refracting telescope by Cuff, London,
with rayskin covered outer tube and vellum covered draw-tubes. This telescope is believed
to have been used by Admiral Rodney at the Battle of the Saints, 1782.*

18th century reflecting telescope of Gregorian type with geared altazimuth adjustment

18th century brass altazimuth theodolite, signed Heath and Wing. *The telescope has a sight to either end and a bubble level to the side.*

17th century universal pocket sundial by Pierre Nory, Gisors, dated 1644, mounted with lunar volvelle with adjustable string gnomon, hour scales and table of latitudes

Late 18th century Dutch brass sextant, frame signed G. Hulst van Keulen Fecit Amsterdam

The Milky Way, the galaxy containing the sun and the earth, contains millions of stars; it is in the approximate shape of a vast flat disc with the solar system somewhere near the central plane and around three fifths of the radius from the centre. To the naked eye on a clear night the Milky Way appears as a luminous band stretching across the heavens in a gigantic arc.

Galileo, Galilei (1564–1642)

Italian astronomer and experimental philosopher, born at Pisa. His father, Vincenzio, was an impoverished descendant of a noble Florentine family who had changed their name from Bonajuti. The young boy from early days displayed an inventive and free thinking mind, occupying many hours constructing toy machines. His education was partly at Vallombrosa monastery and from 1581 at the University of Pisa where he first studied medicine and then switched to experimental philosophy working under Guido Ubaldi who helped him to achieve a mathematical lectureship in Pisa under the Grand Duke Ferdinand I de' Medici.

Observation, patience, perseverance and courage marked his passage through the turbulent intellectual life around him. The facets of his studying were legion. He was a lover of music and painting and yet, parallel with this, was to unleash upon the world of learning a spate of discovery and invention that has only been equalled or even approached by few. The great mass of his notes and papers, when they appeared in a 20 volume edition 250 years after his death, astounded scholars.

His story probably starts with the observation of a swinging bronze lamp in the

Some of the instruments used by Galileo in his study of the heavens

cathedral at Pisa. While pondering this he found the isochronism of pendulum oscillations; that the range of the swing or oscillation had no effect on its duration. He tested his findings by counting the beats of his pulse and comparing the number of pulsations with the time of the pendulum vibration and saw a possibility for using the discovery for chronometers. Early in his life he wrote a paper on the specific gravity of solid bodies and from this idea constructed a hydrostatic balance. This was followed by his discovery that the velocities of all falling bodies great and small were equal; previously educated thought had taken it that a body six times as heavy as another would fall through space in a sixth of the time. He attempted to prove this to people by an experiment from the Leaning Tower of Pisa, but even such a practical demonstration failed to convince many disbelievers. In 1592 he accepted the chair of mathematics at Padua, a relief and retreat for him away from the growing number of enemies he had made in the scholastic circles of Pisa. He worked for 18 years in his new surroundings and the inventions that came here included a proportional compass, a type of thermometer and the all important telescope. Claims were made in Holland that others there had preceded him with the discovery of the principle, notably Lippershey who had by accident placed two convex lenses in line and noticed the magnification of a distant object. It is probable that Galileo realized the full significance of this. He applied the theory of refraction to the problem and within 24 hours it is recorded that he had made his telescope. It had a convex object lens and a concave eyepiece. As he swung this simple instrument towards the heavens the first sightings would have been as rays of light of fresh knowledge – the scarred surface of the moon was revealed, and that it was illuminated by reflected light from the sun; Jupiter had four satellites; Saturn had a triple aspect; that the perfect respected sun was speckled with sunspots and the Milky Way was as Milton, who had visited Galileo, declared, 'powdered with stars'. Demands came in from many places for models of his telescope and these he fulfilled to the best of his ability.

After 1610 he worked in Florence under the patronage of Ferdinand II. Here he was to involve himself in theological controversy. He tried to explain how certain Biblical passages should be read in the light of his theories. Needless to say his ideas ran against accepted thinking and by 1616 he had been ordered by the Inquisition not to make statements that seemed to contradict the Scriptures. In 1632 he published *Dialogue on the Ptolemaic and Copernican Systems* which he dedicated to Ferdinand II. The issue of this treatise was taken up as a direct challenge by the Roman authorities and in particular by a former patron, Pope Urban VIII. Called to Rome he was ordered to refrain from his heresies – in particular to recant on the publicly stated theory on the diurnal and yearly motion of the earth and of the stability of the sun. Authority decreed that he must recite the seven Penitential Psalms once a week and that he was to be held almost a prisoner for the rest of his life. But the freedom of his thinking and discoveries was not something that could be bound by static orders, it had already escaped into the world and there was no way it could again be called back. Galileo's conquests paved the road for those who would follow.

Galvanometer
An instrument for detecting and measuring electric currents.

Gas Chromatograph
A sensitive device for analysing the components of a complex mixture of volatile substances.

Gas Thermometer
An apparatus for measuring temperature by

Thomson's galvanometer, signed
J.Carpentier

observing the pressure of gas at constant volume.

Gauss, Karl Friedrich (1777–1855)

German mathematician born at Brunswick and educated by the patronage of a nobleman who had heard of his early talents. When only 25 he published *Disquisitiones Arithmeticae*, a work which was to bring him considerable acclaim. In 1807 he was appointed director of the Göttingen Observ-

atory, a post he was to keep for the rest of his life. He researched into theoretical astronomy, electricity, magnetism, mechanics and optics. His calculation of the elements of the newly discovered planet Ceres was one of his most important works. His publications included *Intensitas vis magneticae terrestris ad mensuram absolutam revocata*; his collected works were brought out by the Royal Society in seven volumes between 1863 and 1871.

Geissler, Heinrich (1814–79)

German physicist born at Igelshieb in Saxe-Meiningen who was brought up as a glass-blower. He moved to Bonn in 1854 and soon built up a reputation as a skilled producer of chemical apparatus. His research included an investigation of the density of water, also the expansion between ice and the freezing of water. His inventions included the sealed glass tubes, which bear his name, that exhibit the phenomenon that goes with the discharge of electricity through vapours and gases. Items produced by his laboratory included a vaporimeter, mercury air-pump, thermometer and balances.

Geminus Dial *see* Rojas Dial

Generator *see* Dynamo

Geodesy

The science of surveying huge tracts of country on a very large scale, not only for the purpose of producing very accurate maps but also for investigating the curvature of the earth's form and also the figure and the dimensions of the earth. The determination of the earth's dimensions has fascinated scholars from early times. Eratosthenes (*q.v.*) assumed that the earth was spherical and, if so, the dimensions of this sphere could be calculated by measuring the shortest distance between two places. He took Syene and Alexandria as the two places and assumed that they were on the same meridian. The altitude of the sun at Alexandria was found by the gnomon at midsummer, then working with this and the distance between the two places the length of the earth's circumference was calculated with surprising accuracy.

Gerard of Cremona (c.1114–87)

Born at Cremona in Italy but spent most of his life in Toledo in Spain. Here he studied in Spanish Moslem schools and achieved a considerable knowledge of Arabic and then devoted a great deal of his time to translating Arabic and other works into Latin. The most notable example was his translation of Ptolemy's (*q.v.*) *Almagest*; others were the *Tables of Arzakhel* of Toledo, Euclid's *Geometry*, Al Farghani's *Elements of Astronomy* and treatises on algebra, arithmetic and astrology.

Gill, Sir David (1843–1914)

Scottish astronomer born at Blairythan in Aberdeenshire who was educated at Marischal College and the University there. He was the director of the private observatory belonging to Lord Lindsay between 1873 and 1876 and went with his Lordship on an expedition to the Indian Ocean. His researches included the determination of the solar parallax by a study of the movements of Mars. He was the official astronomer at the Cape of Good Hope between 1879 and 1906 and whilst there he carried out the immense work of cataloguing the stars of the southern hemisphere; the sum total of entries came to nearly half a million stars.

Girandole Clock

One of the clock designs by Lemuel Curtis first made around 1815. A wall clock often about 3 ft 6 in high with classical motifs incorporated in the case.

Globe Dial

Spherical equinoctial dial sometimes made of stone; the hour circles are engraved on the sphere.

Gnomon

The Greek word for the style or arm of a sundial that projects the shadow which gives the reading from the dial.

Goniometer

An instrument for the measurement of angles, specifically those of crystals. There are a number of versions of the device: contact goniometer, vertical-circle goniometer, reflecting goniometer and horizontal goniometer. Nicolaus Stena in 1669 was one of the first to seriously study the angles of crystals; the contact goniometer was produced by Carangeot in 1783.

Gothic Clocks

An all-embracing term for certain clocks made in Germany, Italy and Switzerland during the 16th and 17th centuries. The case designs usually leant heavily on Gothic architectural motifs.

Goniometer by Yeates, Dublin, mid-19th century

Governor

Something that can regulate the speed of an engine or clockwork movement, either by fuel restriction or by an expanding spinning device that comes up against some form of braking.

Grandfather Clock

Another name for a longcase clock. One suggestion is that the name possibly derived from the popular music hall song 'My Grandfather's Clock' that was current in Britain and America during the 1880s. Grandmother clocks, small longcase clocks generally not more than 3 ft 6 in high, are rare, and have been 'faked up' to satisfy the demands of fashion by cutting the true longcase clocks.

Graphometer

It may also be known as a Goniometer (*see* above) but goniometer strictly applies only

18th century French brass graphometer by Clerget, Paris, inscribed Clerget Paris au Butterfield

Late 17th century Dutch brass graphometer with central magnetic compass, engraved double 180° scale, signed Anton Sneewins

to an instrument intended for measuring the angles of crystals. The graphometer is a mathematical device for measuring angles in surveying work. In 1597 Danfrie published a description of a kind of graphometer which was really a circumferentor with a semi-circular scale instead of a complete circle. The graphometer was an instrument that could be quite easily made and was uncomplicated to use in the field. Models were still being produced in the early part of the 19th century, although by that time quite advanced theodolites were available. The semi-circular scale could be provided with fixed and movable sights and it could also be detached for use as a protractor on the drawing board.

Greenwich Observatory

The original Royal Observatory was designed by Sir Christopher Wren in 1675; the Observatory stands on the internationally accepted prime meridian of longitude, agreed in 1884 and the basis of Greenwich Mean Time. A new building was completed in 1899, with the magnetic pavilion placed some 400 yards to the east to avoid iron used in the construction of the main building disturbing the instruments.

Gregorian Calendar

The revision of the Julian calendar introduced by Pope Gregory XIII, which is still in force. The year consists of 365 days with a leap year of 366 days in every year whose number is divisible by four; the only exceptions are centenary years which are not divisible by 400, such as 1900.

Gregorian Telescope

A reflecting instrument with which light is thrown on to a second smaller mirror at the upper end of the tube and thence back to an eyepiece mounted behind a hole in the centre of the principal mirror. It was developed by James Gregory (*q.v.*) of Aberdeen. He published *Optica Promota* in 1663 in which he proposed the making of the first reflecting telescope with two concave mirrors. The large collector would be paraboloidal with the smaller ellipsoidal which would focus the image at the centre of the main speculum in which there would be a small hole to allow the light to pass to the eyepiece. James Gregory got as far as commissioning two London opticians to make the mirrors but, finding their efforts unsatisfactory, he dropped the idea. However, the plan was worked on by John Hadley (*q.v.*) who produced an instrument in 1731 that acted successfully. During the 18th century many beautiful Gregorians were made by the celebrated Scottish optician, James Short of Edinburgh. It seems that in 1674 Robert Hooke (*q.v.*) also set out to make a telescope to this plan but was not too successful.

Gregory Family

A Scottish family, a number of whose members contributed to the progress of the sciences.

David Gregory (1627–1717) eldest son of the Rev John Gregory of Drumoak, Aberdeenshire. For some time he was occupied with a business in Holland but when he succeeded to the family seat he came back to Scotland and spent his time in scientific pursuits. He is reputed to have been the first person in the north of Scotland to own a barometer. This enabled him to make weather predictions, which promptly brought down on his head accusations of witchcraft from the presbytery. After some wrangling he was able to clear himself.

James Gregory (1638–75) mathematician and younger brother of David (*q.v.*). He was educated at the grammar school in Aberdeen and then Marischal College. Around 1665 he went to the University of Padua, and two years later wrote *Vera circuli et hyperbolae quadratura* – a treatise mainly concerned with the circle and hyperbola. In 1669 he was appointed professor of mathematics at the University of St Andrews and in 1674 moved to the chair in the same subject in Edinburgh. In 1675 he was the victim of tragedy, being struck with blindness whilst demonstrating to his students the observation of Jupiter through one of his telescopes.

David Gregory (1661–1708) son of David Gregory (*q.v.*), educated for some time in Aberdeen and then Edinburgh. In 1683 he was appointed professor of mathematics. Eight years later he moved to Oxford to become the Savilian professor of astronomy. His published works included *Astronomiae physicae et geometricae elementa*, a work respected by Sir Isaac Newton (*q.v.*), and *A Treatise on Practical Geometry*.

John Gregory (1724–73) grandson of James Gregory (*q.v.*), the mathematician, he studied medicine and held leading posts in

London, Aberdeen and Edinburgh.

James Gregory (1753–1821) eldest son of John (*q.v.*), he also studied medicine and finished his career as President of the Edinburgh College of Physicians.

William Gregory (1803–58) son of James (*q.v.*), he became the professor of chemistry at the Andersonian Institution, Glasgow, then at Aberdeen and then Edinburgh.

Duncan Farquharson Gregory (1813–44) brother of William (*q.v.*), he studied mathematics and chemistry and for a time was assistant professor of chemistry at Trinity College, Cambridge. He started the *Cambridge Mathematical Journal* and wrote a treatise on the *Application of Analysis to Solid Geometry*.

Groma

One of the earliest devices for setting lines at right angles; basically it consisted of two centre ribs lashed together at right angles. At the ends of each rib were plumb lines held vertical by some kind of weight; there was also a loop of cord for suspending the groma by hand. Hero of Alexandria mentioned a device called the Grecian Star used by his surveyors; the Romans would have also used some form of instrument like a groma but of an improved and more accurate design. As centuries passed the groma was rendered obsolete by the surveyor's cross and the optical square.

Gunner's Quadrant

An instrument consisting of a graduated limb, with a spirit level, and an arm by which it is applied to a cannon or mortar in adjusting the piece to the elevation for the desired range.

Gunter, Edmund (1581–1626)

English mathematician of Welsh extraction born in Herefordshire, he was educated at Westminster School and elected a student of Christ Church, Oxford. He took orders to become a preacher but an underlying enthusiasm for mathematics and astronomy resulted in his appointment as professor of astronomy at Gresham College, London. A number of worthy inventions are attributed to Gunter including the sector, bow and quadrant. He also researched into magnetism as applied to a compass needle. His practical legacy to scientists, sailors and others is Gunter's Chain, for surveying. This is 22 yards long and has 100 links; his line, used with logarithms, enabled the solution of problems by instrument; his Quadrant is used to find the hour, the sun's azimuth and also to take the altitude of an object and his scale, by which, with the aid of a pair of compasses, problems of navigation could be solved. By command of James I in 1624 he published *The Description and Use of His Majestie's Dials in Whitehall Garden*.

Gyro Compass

One that does not make use of magnetism but uses a motor driven gyroscope to indicate true north.

Gyroscope

A spinning wheel, so mounted that it can rotate about any axis.

Gyrostat

Designed by Lord Kelvin (*q.v.*), this is more intricate than the gyroscope. It was intended to illustrate the more complicated state of motion of a spinning body when free to wander about on a horizontal plane.

Habermel, Erasmus (active late 16th century)

Celebrated instrument maker from Prague whose workshop was a model for many apprentices. He made a number of instruments for his patrons including astrolabes and surveyor's and gunner's triangulation devices.

Hadley, John (1682–1744)

English mathematician and mechanician who is principally recalled for the great work that he did to improve the reflecting telescope. Navigators owe him a greater debt for his invention of the reflecting quadrant. The invention was actually challenged by Thomas Godfrey of Philadelphia who claimed that he had been the first, but it was proved that both men had been working independently and without the knowledge of the other's advances. The basis of both models was the observation of the sun or a star by double reflection; there was a fixed or horizon mirror on the left-hand arm which was half clear and half silvered. The seaman looked through the peephole sight and saw the horizon through the clear glass, he then adjusted the index arm until the sun or star appeared reflected by both mirrors; the elevation of the heavenly body could be read from a scale on the arc.

Hale, George Ellery (1868–1938)

American astronomer from Chicago who graduated from the Massachusetts Institute of Technology in 1890. His astronomical research was conducted at the Harvard Observatory and at Berlin University. When he returned to his native city he organized the Kenwood Observatory. In 1892 he was made professor of astronomy at the University of Chicago and was instrumental in the founding of both Yerkes and Mount Palomar Observatories. He invented an instrument called a spectroheliograph for studying the sun, he discovered the existence of magnetic fields in sunspots and of the periodic reversal in the polarizing of these magnetic fields. His written works included *The Story of Solar Evolution, The Depth of the Universe* and *Beyond the Milky Way*.

Halley, Edmund (1656–1742)

English astronomer born at Haggerston, London, the son of a wealthy soapboiler. Edmund attended St Paul's school and later Queen's College, Oxford. In 1676 he went to St Helena to observe the stars and earned himself the nickname 'Southern Tycho'. His particular studies included the moon, gravity, the variation of the compass in the Atlantic, Jupiter and Saturn, proper motion of fixed stars, possible magnetic origin of the aurora borealis and comets – in particular that one sighted in 1682 which was the same one that had appeared in 1456, 1531 and 1607 and which Halley predicted would reappear again in 1758. It now bears his name.

Halo

A luminous ring surrounding the sun or moon at times, it is caused by the refraction of light by ice crystals in the atmosphere. The word is derived from the Greek for 'threshing floor' because of the circular path walked by the oxen when the corn is being threshed. The theory for explaining the formation of Halos was first presented by René Descartes (1596–1650) and it was adopted by Sir Isaac Newton (*q.v.*). The validity of the theory was clearly demonstrated in a memoir published in the *Journal de l'École* [Royale Polytechnique] for 1847 (XVIII, 1–270) by A. Bravais.

Harcourt Pentane Lamp

One that burns with a steady flame using pentane. It has been considered as a standard for a light source.

Harrison, John (1693–1776)

English horologist born at Faulby, near Pontefract in Yorkshire, the son of a carpenter. At first John followed his father's trade and made some pocket money doing surveying and land measuring but the fascination for mechanical things took over and

he began to concentrate on timepieces. In 1715 he produced a clock with wooden wheels, which predated the wooden movements of Silas Hoadley of Connecticut by more than 100 years. By 1726 he had worked out the clever 'gridiron pendulum' which could keep its correct length by compensating for alterations of temperature. He rose to the challenge of the rewards for producing an accurate marine timekeeper and by 1728 he had prepared designs for such an instrument. In 1735 he went to the authorities with a timepiece which was also seen by Edmund Halley (*q.v.*) whose influence assisted him in taking passage to Lisbon to test the piece. The first results were so encouraging that he was granted £500 to continue working on it. A persevering craftsman he made model after model until in 1761 he was satisfied and his son William was sent to Jamaica in HMS *Deptford* to test it; on the return to Portsmouth the chronometer was found to have lost just one minute and 54½ seconds.

Hartman, Georg (1489–1564)

Instrument maker from Nuremberg, one of the first skilled craftsmen of his time to sign his work.

Hauksbee, Francis (d.c.1713)

English scientist who was among the first to experiment with electricity. He found that what could be termed electric light could be produced by rubbing together amber, glass and wool. He made a simple device with a glass cylinder that when stroked by hand produced a current. An improved air-pump bears his name.

Heaven

To the ancients, that domed expanse above the earth in which the sun, moon, planets and stars seem to be placed. In the cosmogonies of the early peoples there is often a plurality of heavens, between three and seven, the higher transcending the lower in glory.

Heliacal

The rising of a celestial object, a star, at the same time as the rising of the sun, so that it could be seen emerging from the rays of the sun. The term could also be applied to the setting of the star as sight of it was lost by nearness to the sun.

Heliocentric

Having the sun as a centre; relative to the sun.

Heliograph

An instrument that can be used for signalling between two distant points. It relies on flashing the sun's rays by reflecting them from a small mirror mounted on a stand that can be moved quickly to produce flashes of short and long duration, as in Morse code. The device can be aimed by means of sights.

Heliometer

A refracting telescope which has a split objective lens; its main purposes are to determine small angular distances between celestial bodies and to measure their diameters. The instrument was invented in 1814 by Fraunhofer (q.v.) and as the name implies was first intended to take solar measurements.

Heliostat

An instrument that will reflect the rays of the sun in a fixed and constant direction regardless of the motion of the sun. The optical device normally consists of a mirror which is mounted on an axis parallel to the axis of the earth and which is rotated with the same angular velocity as the sun. One of the earliest mentions of a heliostat is described by Wilhelm Jacob in the third edition of his *Physices elementa* published in 1742.

Henry, Joseph (1797–1878)

American physicist born at Albany, New York and educated at the Albany Academy. He showed early talent for research into electricity and can be credited with improvements to the electromagnet and the invention of a type of electric telegraph. He did not patent his idea and in 1844, some 13 years later, Samuel Morse (q.v.) took the credit for the first workable electric telegraph.

Hercules

A constellation of the northern hemisphere noted by Eudoxus (q.v.) and Aratus and catalogued by Ptolemy (q.v.) as having 29 stars and later by Tycho Brahe (q.v.) who made the total 28. Herculis, a coloured double star in this constellation, was discovered by Sir William Herschel (q.v.) in 1782.

Herodotus (c.484–c.425 BC)

The great Greek historian, popularly termed the 'Father of History' was born at Halicarnassus in Asia Minor. He had the essential curious mind that would inquire into almost any subject. He mentioned the eclipse of 585 BC telling us that it came during a battle in the war which had been raging between the Lydians and the Medes for years; the sudden darkness stopped the fighting and a peace was arranged. Herodotus adds that the eclipse had been foretold by Thales (*q.v.*), who was considered the founder of Greek astronomy.

Herschel, Sir John Frederick William (1792–1871)

English astronomer born at Slough, Buckinghamshire, the son of Sir William Herschel (*q.v.*). He was educated at Eton and St John's College, Cambridge. At first it looked as though he might enter the legal profession but in 1816 he gave up other matters for astronomy. In 1822 he began a systematic study of the heavens, surveying the whole of the northern hemisphere, identifying more than 500 nebulae and nearly 4,000 double stars. In 1834 he established an observatory at Feldhausen near Cape Town where he spent four years observing the southern hemisphere. In 1847 he published *Results of Astronomical Observations made at the Cape of Good Hope*.

Herschel, Lucretia Caroline (1750–1848)

Anglo-German astronomer born at Hanover and the sister of Sir William Herschel (*q.v.*), she came to England in 1772 and became an assistant to her brother. She was a dedicated worker and proved of great assistance with research and tedious but necessary observations. When her brother became the private astronomer to George III she remained helping him and was awarded an annual salary by the King of £50. For personal amusement she would quarter the heavens with a small Newtonian telescope, and by this means she discovered three remarkable nebulae and eight comets, five of them with unquestionable priority. She also added 561 stars to the catalogue published by Flamsteed (*q.v.*). She was elected as an honorary member of the Astronomical Society from whom she also received a gold medal for her work.

Herschel, Sir William (1738–1822)

Anglo-German astronomer born at Hanover, the son of an oboe player in the Hanoverian guard. His education was piecemeal owing to unrest in the country and he was largely self taught in science and astronomy, although trained as a professional musician. He came to England in 1757 and for a time taught music in Leeds and other north country towns and then in 1766 he was appointed organist at the Octagon Chapel in Bath. In this city he turned his attention to astronomy and using a telescope he had made himself be began to observe the heavens. In 1781 Sir William discovered a new planet called at first Georgium Sidus and later Uranus; the following year he received the appointment of private astronomer to George III and moved to Slough. Further discoveries included two satellites of Saturn, the phenomenon of the motion of the double stars round each other, the rotation of Saturn and Venus and important aspects of the Milky Way. Whilst at Slough he wrote his famous *Motion of the Solar System*, which was published in 1783. Six years later he erected his great telescope, which had a 40 ft focal length and a 4 ft aperture. With this, at first viewing, he saw Saturn complete with satellites with a new fresh brightness, although, as it was later remarked, it was strange that he should have missed the eighth satellite, Hyperion.

Hevelius, Johann (1611–87)

The name is also spelt Hevel or Höwelcke. German astronomer born at Danzig, he studied at Leiden and then after travelling to France and England he settled down in his native city and became a brewer. In 1639 his interest in astronomy was aroused and two years later he built a well stocked observatory, with fine instruments including a monster tubeless aerial or skeletal telescope with a focal length of 150 ft in his house. From an illustration in his *Machina Coelestis* this amazing instrument was hauled up a tall pole by a block and tackle which could be used to give it elevation and to hold the lenses in line. The whole contraption dwarfed a nearby coach and horses. Honoured by a visit from John II and Maria Gonzaga, King and Queen of Poland, Hevelius worked hard and produced such results as the following: four comets, accurate observation of sunspots, charting of the surface of the moon and found the moon's libration in longitude. On 26 September 1679 his observatory, his beloved instruments and books were destroyed by fire, the act of malicious jealousy. The brutal shock of this affected his strength and, although he did repair the observatory sufficiently to see the great comet of December 1680, the lust for exploration of the great heavens had weakened. His written works included *Seleno-graphia, Annus climactericus, Prodromus cometicus, Cometographia, Firmamentum Sobiescianum* and *Prodromus astronomiae* – his catalogue of 1,564 stars which appeared posthumously.

Hipparchus (c.160–120 BC)

Greek astronomer born at Nicea in Bithynia, he observed for lengthy periods from Rhodes and can be justly claimed to be the founder of scientific astronomy. He discovered the precession of the equinoxes, his catalogue of stars numbered 850 and he showed amongst other matters that places of the globe could be more accurately located by reference to latitude and longitude.

Holcomb, Amasa (1787–1875)

American telescope maker born at Granby, Connecticut. His early education was sketchy until the age of 15 when he passed the entrance examination for a school in Suffield, Connecticut. In his free time he was able to study a collection of books left by his uncle Abijah Holcomb; amongst these were volumes on astronomy, geometry and navigation. He started making simple instruments and with one of these he was able to observe the eclipse of June 1806. As time passed he became more skilful and was soon able to offer compasses, chains, scales, protractors and dividers; he was also making levelling instruments, electrical machines and magnets. By 1830 he had completed an achromatic telescope; his reflecting telescopes were nearly always of the Herschelian type.

Holometer

Also termed a pantometer, it is used for

measuring all angles in determination of elevations and distances.

Hooke, Robert (1635–1703)

English experimental philosopher born at Freshwater in the Isle of Wight, and educated at Westminster School and later Christ Church, Oxford. For a short time he assisted Robert Boyle (*q.v.*) and then was appointed as curator of experiments to the Royal Society. Among his inventions were a double-barrelled air-pump, a spirit level aerometer (an instrument for determining the density of gases), a marine barometer and a sea gauge. Robert Hooke was a strange man, deeply religious and highly moral, ragged tempered and solitary. He had a stooped figure with shrunken limbs and long dishevelled locks. He quarrelled with Newton (*q.v.*) about aspects of the discovery of gravity, saying that much of it was his earlier work; there was also a long running argument with Johann Hevelius (*q.v.*) over telescopic sights. His written works were *Lectiones Cutlerianae* and *Micrographia*.

Hooke's Microscope

A compound microscope with the main body about 300 mm long made of cardboard covered with gold tooled dark red leather. The eye lens was bi-convex and mounted in a wooden holder with a dust-cap to hold spare objectives. Mounted internally was a plano-convex field lens. Focusing was done by twisting the bottom lens holder.

Horizontal Dial

The normal garden sundial; one in which the hour-scale is horizontal.

Horology

The science of the making of timepieces.

Hughes, David Edward (1831–1900)

Anglo-American electrician, born in London and went to Virginia in 1837. He became professor of music at Bardstown College, Kentucky. Devices he invented or developed were an improved telegraph type-printer and a microphone which was similar to that produced at the same time by Lüdtge.

Huygens, Christiaan (1629–95)

Dutch mathematician, astronomer and physicist born in The Hague. He was educated at home by his father, then he went to Leiden and finally the juridical school at Breda. He travelled through his home country, Denmark, France and England. He worked on improvements in telescope design and in particular on lenses. In 1655 he discovered a satellite of Saturn and four years later the ring of Saturn. He was one of the first to think of using a circular pendulum with clocks. He concentrated on making lenses with very large focal lengths – up to 210 ft. His *Horologium oscillatorium*, published in 1673, was dedicated to Louis XIV.

Hydraulic Press

Invented by Joseph Bramah (*q.v.*) in 1785, it incorporates the well known hydrostatic principle that the pressure on any part of the surface of any liquid is transmitted equally in all directions. This can be demonstrated using two connected cylinders, one comparatively small with a small piston, the other as large as needed for the particular problem. By means of this simple device very large forces can be used.

Hydrography

A scientific study of the waters of the globe; the surveying and mapping of oceans, seas and rivers.

Hydrometer

An instrument for finding the densities of liquids.

Hygrometer

A device for measuring the relative humidity of the air and its moisture content.

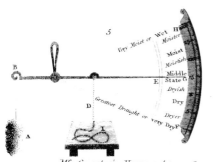

Mr. Coventry's Hygrometer.   7

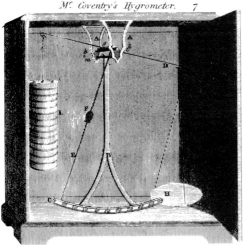

John Coventry's hygrometer, c. 1810

Hypsography

The study and mapping of the earth's topography above sea level.

Hypsometer

An instrument for determining altitudes by calculating the boiling point of water at different heights above sea level.

Immersion Objective Lens

One that allows more light to enter the optical system of the microscope. This is done by placing a drop of cedarwood oil directly on the slide to be examined; the lowest lens of the objectives is then brought down until it is in the oil. The technique is reserved for high power instruments.

Inclination

The degree of deviation from a horizontal or vertical plane.

Inclining Dial

One which allows the tilting of the hour-plate from the horizontal plane to different latitudes. This really transforms it into a universal dial.

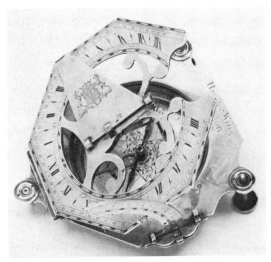

18th century silver inclining dial by Heath and Wing, London; base inset with compass and levels, hinged gnomon

Inclinometer

An instrument to determine magnetic dip. It is used to show the attitude of a ship relative to the horizontal.

Induction Coil

A type of transformer for producing high voltage from a low voltage.

Innes, Robert Thornton Axton (1861–1933)

Scottish astronomer born at Edinburgh. As director of the Transvaal Observatory he discovered a large number of double stars in the southern hemisphere and became one of the greatest authorities on this phenomenon.

Ives, Frederic Eugene (1856–1937)

American inventor who became manager of the Cornell University photographic laboratory. His accomplishments included a half tone printing process, in which the tones of the picture were reduced to small dots of varying sizes. To this he added a process for use with gravure with which the ink was carried not by a system of small dots but by small pits etched into the plate, strictly an intaglio method. He also developed a photo-chromoscope, a binocular microscope and an instrument for optically reproducing objects in natural colour.

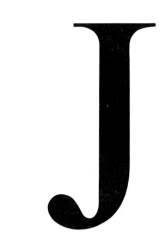

Jacobi, Karl Gustav Jacob (1804–51)

German mathematician born at Potsdam who, after completing his education at Berlin University, became professor of mathematics at Königsberg. His principal research was into elliptic functions. His most important work is *Fundamenta Nova Theoriae Functionum Ellipticarum*.

Jacob's Staff

A short square rod with a cursor, which is used for measuring heights and distances. It can also be a support for a compass.

Jacob's staff

Jones, Henry (1664–95)

English clockmaker born at Boldre, Southampton, the son of the local vicar, and apprenticed to Edward East (1627–97). He was a most respected maker and had a reputation for never producing poor timepieces. There is a story that one of Henry's finest clocks, made for Charles II, was later given to one of his lady friends. She subsequently passed it to Robert Seignor of Exchange Alley to be repaired, who in his turn passed it to another clockmaker, Edward Staunton to erase the engraved

'Henricus Jones, Londini' and to replace this with his own mark.

Joule, James Prescott (1818–89)

English physicist born at Salford near Manchester. Although he had some schooling from John Dalton (*q.v.*), particularly in chemistry, he taught himself electricity and electromagnetism. He had an impelling desire for great accuracy and thus he moved towards establishing a scientific unit for work in practical electricity. This is taken to be the work done in one second by a current of one ampere flowing through a resistance of one ohm – the unit has since carried his name. He formulated the first law of thermodynamics, the law of conservation of energy. This states that energy may be converted from one form to another but cannot be created or destroyed. In *Annals of Electricity*, published in 1837, Joule detailed an electromagnetic engine which he had developed. His scientific papers were collected and published by the Physical Society of London. The first volume, brought out in 1884, contained details of the work he had been personally responsible for, and the second, brought out in 1887, was an account of research carried out with other scholars.

Julian Calendar

Introduced in Rome in 46 BC by Julius Caesar. It differed from the one in use today only in two points: the start of the year was not fixed as being on 1 January and, as well as occurring every fourth year, the leap years came up in every centenary year. It was replaced by the Gregorian Calendar in 1582.

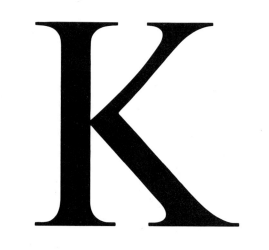

Kaleidoscope

An optical instrument developed by Sir David Brewster (*q.v.*) in 1817: it consists of a tube about 1 ft long, inside which are two, sometimes three, mirrors that extend for the whole length of the tube and are placed at an angle of 60 degrees. One end of the tube is closed by a plate with a small hole or eyeglass; the other end is closed by a fixed plate of clear glass on which are placed small fragments of coloured glass. As the tube is revolved these fragments alter their positions and give rise to colourful and changing patterns. With a polyangular kaleidoscope the angle between the mirrors can be changed at will. The basic idea of the kaleidoscope had been experimented with 100 years before by R. Bradley. Although really a toy, the instrument can be used to teach students the formation of multiple images.

Kapp, Gisbert (1842–1922)

Austrian engineer born at Mauer near Vienna, the son of a native of Trieste and his Scottish wife. He studied in Zurich. In 1875 he was working in England and later he went to Berlin in 1894, returning to England in 1904. His accomplishments included a theory on electric currents, a dynamo and improvements to electrical testing methods.

Kapteyn, Jacobus Cornelius (1851–1922)

Dutch astronomer who carried out valuable research on stellar parallaxes which was to lead to advances in the knowledge of the structure of the universe. He also collaborated with Sir David Gill (*q.v.*) on preparing a star catalogue of the southern hemisphere, working on calculations and measurements from photographs taken by his partner.

Kater, Henry (1777–1835)

English physicist of German descent who was born at Bristol. He joined the army and during service in India gave valuable assistance with the trigonometrical survey. Ill health forced him to retire from the army and he devoted himself to scientific pursuits. He pointed to the superiority of the Cassegrain telescope over the Gregorian, stating that the illuminating power of the former was greater. He invented a floating collimator, a type of small telescope that can be attached to a larger instrument to assist in lining up an object to be observed. He also conducted experiments to determine the length of a seconds' pendulum. In 1814 he was decorated with the Order of St Anne

Brass kaleidoscope with oil illuminant, glass chimney and metal reflector; English, second quarter of 19th century

by the Russians for his work in verifying their standards of length.

Kelvin, Lord William Thomson, Baron Kelvin of Largs (1824–1907)

Scottish physicist born in Belfast but lived in Glasgow after 1832. His education included periods in Glasgow, Cambridge and Paris. He was interested in the work of Joule (*q.v.*) and through his own researches developed the absolute temperature scale and worked on the theory of thermodynamics. He also experimented with electricity, heat, magnetism, elasticity and vortexmotion and his inventions included the mirror galvanometer. He played a leading part in the laying of the first Atlantic cable and he evolved the meter which records domestic consumption of electricity. During his life he published more than 300 papers on scientific subjects.

Kepler, Johann (1571–1630)

German astronomer born at Weil in the duchy of Württemberg. His father was a dissolute character who preferred soldiering with the Duke of Alva to the responsibilities of family life. He later took to tavern keeping and then deserted his home. Johann had also to compete with a raggle-taggle mother and early childhood physical troubles that left him with impaired eyesight and crippled hands. His first education was at a school in Elmendingen followed by a time at the monastic school of Maulbronn which was paid for by the Duke of Württemberg. His brilliance enabled him to enter the University of Tübingen and there he was ably instructed in the principles of Copernicus (*q.v.*) by Michael

Maestlin, who was to remain a lifelong friend. In 1594 he took the chair of astronomy at Grätz. Here he found that he was supposed to supply information to the publishers of almanacs and thus he had to burrow into the astrological ideas of Ptolemy (*q.v.*) and Cardan (*q.v.*).

In 1600 Tycho Brahe (*q.v.*) offered him the position of his assistant at his observatory near Prague. After Brahe's death the Emperor Rudolph II appointed Kepler to the post of imperial mathematician and gave him access to Brahe's observations. His life still remained rugged, he was persecuted for religious beliefs he did not hold but was convinced that there was a harmony in the relationships of the planets. In 1611 his wife and favourite child died and he found himself owed large amounts of salary. His mother, Catherine, through ill-advised talk and thought, got herself arrested on a formal charge of witchcraft and it took all the exertions of her son to procure her release, after 13 months of imprisonment

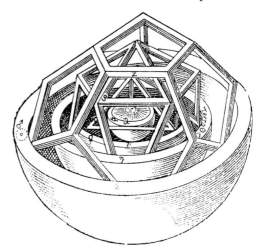

Kepler's system of regular solids

during which she had been exposed to the horrible ordeal of 'territion', the examination under threat of torture; she survived her freedom by only a few months.

Still the great mind of Johann continued upwards, seemingly unimpaired by the road he was treading. As an astronomer he was working towards what he saw could be the establishment of a new foundation and theory for the science. He discovered the three great laws of planetary motion which carry his name, although these were not proved until Newton's (*q.v.*) *Principia* appeared many years later. His extensive literary works were bought by the Empress Catherine II in 1724 from some Frankfurt merchants and for a long time rested in the Observatory of Pulkowa, inaccessible to scholars. They were at last published in eight volumes under the editorship of Dr Frisch between 1858 and 1871.

Kepler's Laws

These are:
1   Planets move round the sun, their orbits being elliptical.
2   The radius joining each planet with the sun describes equal areas in equal times.
3   The squares of the periods of any two planets are proportional to the cubes of their mean distances from the sun.

Kinematics

The branch of the science of mechanics which discusses the phenomena of motion without reference to any force or mass.

Kinetics

The branch of mechanics which discusses the phenomena of motion as affected by force.

Kohlrausch, Friedrich Wilhelm (1840–1910)

German physicist born at Rintelm-am-Weser who conducted considerable research into electromagnetism and invented the Kohlrausch method of investigating electrolytic conductivity.

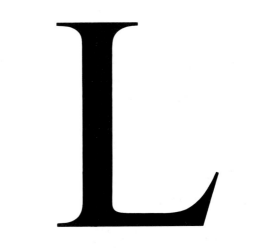

Lalande, Joseph Jérôme de (1732–1807)

French astronomer born at Bourg and sent to Paris to study law. He lodged in the Hôtel Cluny where J. N. Delisle had his observatory and was soon under the thrall of the science of astronomy. In 1751 he was sent to Berlin to make observations on the lunar parallax in conjunction with those being made by Lacaille at the Cape of Good Hope. He worked to improve the planetary theory and in 1759 published a corrected edition of Halley's (*q.v.*) tables. He also wrote *Traité d'Astronomie* and *Histoire Céleste Française*.

Late Regency lancet bracket clock, satinwood with ebony inlay and silvered dial

Lambert, Johann Heinrich (1728–77)

German astronomer, physicist and mathematician born at Mulhausen, Alsace. The son of a tailor, he more or less educated himself. At first he acted as book-keeper at a local ironworks and then became secretary to Professor Iselin who was editing a newspaper in Basle, and then he moved to be private tutor to the children of Count A. von Salis of Coire, where he had the opportunity to study in a first class library. Astronomically, he left a theorem relating to the apparent curvature of the geocentric path of a comet and thoughts on the time of describing an elliptic arc under the Newtonian Law of gravitation. Mathematically, he worked on equations and trigonometry and also made important geometrical discoveries which were published in his *Die freie Perspective*.

Lancet Clock

A type of bracket clock, popular about 1800. It took its name from the early Gothic lancet arch on which the top of its case was modelled.

Lantern Clock

A bracket or wall clock that was weight driven, with a metal frame and a rounded bell dome. An English development it was also called a birdcage clock and sometimes, quite wrongly, a Cromwellian clock.

Latitude

Circles parallel to the Equator which is itself 0 degrees of latitude. The angular distance in degrees north or south of the Equator.

Leeuwenhoek, Anthony van (1632–1723)

Dutch microscopist born at Delft, he made a number of important anatomical discoveries which were published in the *Philosophical Transactions* of the Royal Society in 1680.

Lens

In optics, a piece of glass or other transparent material ground to shape that will diverge or converge transmitted light to form an image.

Leonardo da Vinci (1452–1519)

The great Italian master-genius was born at Vinci and from the earliest years showed promise of what was to come. His father placed him in the studio workshop of Andrea del Verrocchio where Leonardo would have been working alongside Sandro Botticelli and Pietro Perugino and others – a creative atmosphere of painters, sculptors and craftsmen. Leonardo must have had an intellect that brooked no refusal nor baulked at any intellectual challenge regardless, it seems, of subject or magnitude. Walter Pater wrote of him:

Curiosity and the desire of beauty – these are the two elementary forces in Leonardo's genius; curiosity often in conflict with the desire for beauty, but generating, in union with it, a type of subtle and curious grace.

Enlarging on the scope of this man of the Renaissance, Edward McCurdy says,

Anatomist, mathematician, chemist, geologist, botanist, astronomer, geographer – the application of each of these titles is fully justified by the contents of his

Brass lighthouse clock, revolving turret mounted with Centigrade and Fahrenheit thermometers and inset with an aneroid barometer; English, late 19th century

manuscripts at Milan, Paris, Windsor, and London. To estimate aright the value of his researches in the various domains of science would require an almost encyclopedic width of knowledge.

Add to this his skill as a painter, draughtsman in many media, modeller with clay, lute player, and then his amazing ventures into engineering, engines of war and reaching up towards flight. Perhaps his greatest service today is to act as a guide, an encourager by example. A look at Leonardo's sketchbooks, a glance at his writings can be a most liberating experience that can tone up the thought towards a greater assimilation of the work of others.

Leyden Jar

An early type of electrostatic capacitor which basically consists of a thin glass jar with an interior and exterior coating of tin foil. The interior surface is connected to a rod passing through an insulated stopper. If the two metal surfaces are connected for a short time to the terminals of an electromotive force, such as a voltaic battery, induction coil or electric machine, the jar will have the ability to store electric energy that can be recovered as an electrical discharge.

Libration

In astronomy, the oscillation of the moon's face from side to side; the movement is compounded of the moon's rotation on her axis and the orbital motion by which her visible hemisphere changes.

Lighthouse Clock

A shelf clock made by Simon Willard, an American, one of four famous brothers, all

Late 19th century German Wimshurst machine connected with two Leyden jars

clockmakers. Its design was close to that of a lighthouse with a dome-shaped glass hood. It was patented in 1822. Later the term was loosely applied to novelty clocks, in which the case was made not only from metal and glass but also from wood and serpentine.

Lodestone

A variety of natural iron oxide. A number of writers of the past have noted a stone which probably to them had something magical about it. Small pieces of iron would be attracted to it. Just when this magnetism was first used as a navigational aid in a simple compass is not known – myth, legend and imagination confuse the issue. By repute, the Chinese knew of it; Thales of Miletus (q.v.), a merchant of Tyrian parentage, certainly knew of the lodestone, possibly he got his knowledge from the adventurous Phoenicians. The first word that tells of its use in Scandinavia comes around the time of Leif Ericson's voyage to Vinland. Peter Peregrinus writes that if the stone is made to float on water it will turn until its poles point to the poles of the heavens and continues, '...and if this stone be moved aside a thousand times, a thousand times will it return to its place by the will of God.' The lodestone became a collector's curiosity, a must for inclusion in the cabinets of the rich who liked to play with scientific instruments and objects.

Lodge, Sir Oliver Joseph (1851–1940)

English physicist, born at Penkull, Staffordshire, he attended the grammar school at Newport, Salop, and then entered University College, London. His original research and inventions included investigations on lighting, electromotive force in the voltaic cell, electrolysis, electromagnetic waves, wireless telegraphy, electricity and its possible use to disperse fog and smoke. His books included *Life and Matter, Man and the Universe, Ether and Reality, Evolution and Creation, Atoms and Rays, Signalling across Space without Wires* and *Beyond Physics.*

Log

A device consisting of a float and some kind of rotation propeller on a line which could be towed astern of a vessel to record speed and distance covered. The term also means the detailed record of the voyage.

Longitude

The lines of longitude are the great circles of the earth intersecting at the poles. Longitude is measured in degrees east or west of the prime meridian at Greenwich.

Lunar Dial

One for showing the hour of the night by the shadow cast by a gnomon in moonlight.

Lunarium

A mechanical device to show the motion and phases of the moon; it can also demonstrate the occurrence of eclipses.

Lunation

The period of return of the moon to the same position relative to the sun; that is, from full moon back to full moon. The length of this period is 29 days, 12 hours and 45 minutes.

Lyre Clock

Shelf or wall clock whose case roughly represents a lyre. It is probable that the first one was made by one of the sons of Aaron Willard, the American clockmaker who first opened a workshop in Roxbury about 1780 and later moved to Boston, Massachusetts where he opened a small factory.

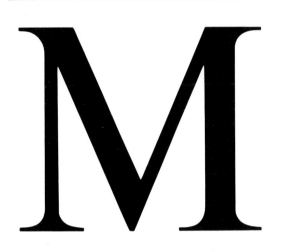

Magellanic Clouds

Two cloudlike patches of nebulous looking light, almost circular in shape and situated in the southern hemisphere. They bear some resemblance to the Milky Way but are not connected to it. They were named by Andrea Corsali after Ferdinand Magellan, the great Portuguese navigator.

Magnetic Azimuth Dial

One in which a magnetic compass needle tells the time as the dial is pointed toward the sun.

Magnetic Compass Dial

Usually a small one with a compass card marked with hour lines; it may also have a gnomon. It was popular in the 18th century and early years of the 19th.

Magnetic Pole

Either of two variable points on the earth's surface towards which a magnetic needle will point; also, either of two areas on a magnet where the magnetic induction is concentrated.

Magnetism, Terrestrial

The science of studying the magnetic phenomena of the earth.

Magnetograph

An instrument for automatically recording the daily variations of either the horizontal or vertical components of the earth's magnetic force. These devices are basically super-sensitive magnetometers.

Magnetometer

An instrument for comparing the strengths of magnetic fields. The usual model consists of a short magnet pivoted so that it can rotate in a horizontal plane. This magnet carries a lightweight non-magnetic pointer fixed at right angles to its centre. The pointer moves over a circular scale marked off in degrees.

Magnitude of Stars

The apparent brightness of a celestial body as found on a numerical scale. On this scale the bright stars have a low reading; thus the sun, taken as the strongest, registers at -27 and the scale goes up to $+23$. Measurement is either by eye in which case it is termed visual magnitude or by a photographic method when it is known as photographic magnitude. It is taken that a star of any one particular magnitude is about 2.51 times brighter than that of the next magnitude. The term or a synonym was understood by Ptolemy ($q.v.$), who published a catalogue of the stars visible in the northern hemisphere. These stars he divided into six grades according to brightness, although size also had some bearing on the division. Ptolemy gave the first magnitude to the brightest star proceeding down the scale to the sixth magnitude for the weakest.

Malmgren, Finn (1895–1928)

Swedish meteorologist born at Falun. He joined up with Amundsen in the North Polar Basin in 1922 and was also meteorologist for three years under Dr Sverdrup. He was severely injured when the airship *Italia* crashed on the ice in 1928 and he sent his companions to safety whilst he remained behind.

Manometer

A device for measuring the pressures exerted by gases or vapours.

Marconi, Marchese Guglielmo (1874–1937)

Italian inventor born at Bologna, the son of Giuseppe and his wife Annie Jameson, an Irish woman. He was educated first at Leghorn under Professor Rosa and then at Bologna University. He made some early experiments with wireless telegraphy at Bologna and then emigrated to England where he was welcomed by Sir William Preece, engineer-in-chief of the Post Office telegraph system and himself a dabbler in experiments. An early successful test was made across the Bristol Channel from Penarth in Wales to Brean Down near Weston-super-Mare. This was followed by a cross Channel success and then in 1901 a signal from Poldhu in Cornwall was received at St John's in Newfoundland; this was followed by communications between Cape Cod, Massachusetts and Cornwall. Marconi's inventive development and the making of radio valves was the genesis of modern radio.

Marine Barometer

An instrument which has a constriction in the bore of the vertical tube to reduce the movement of the level in the tube caused by the motion of the ship in rough weather.

Mariotte, Edmé (c.1620–84)

French mathematician and physicist who was born in Burgundy and spent most of his life at Dijon where he became the prior of St Martin-sous-Beaune. His research and experiments ranged over hydraulics and the composition of air, on which subjects he published a book which contained Boyle's ($q.v.$) law, known in France as Mariotte's law, the nature of colour, the notes of a trumpet, the fall of bodies, barometers, freezing of water and the recoil of guns.

Maudslay, Henry (1771–1831)

English mechanical engineer born at Woolwich. He became a blacksmith at the Arsenal and was also engaged by Bramah (*q.v.*) to produce his inventions. He became one of the finest makers and inventors of machine-tools, including the slide-rest and the first screw cutting device, and he also made radical improvements in the design of lathes.

Maxim, Sir Hiram Stevens (1840–1916)

Anglo-American engineer and inventor born at Sangerville, Maine. After an apprenticeship with a coach builder he worked for a time in his uncle's machine works and then became a draughtsman in the Novelty Ironworks and Shipbuilding Company, New York. His fertile brain conceived, amongst other devices, gas generating plants, steam and vacuum pumps, improvements in electric lamps and the famous Maxim gun. He came to England and was naturalized and knighted in 1901.

Maximum and Minimum Thermometer

One that will register the highest and lowest temperatures in a given period. At each end of the mercury column are tiny metal slivers which are moved by the mercury and then left as the temperature changes, so registering the maximum and minimum for the period.

Mayer, Johann Tobias (1723–62)

German astronomer born at Marbach in Württemberg and raised in poor circumstances. To a large degree he was self taught and entered Homann's cartographic workshop in Nuremberg. Here he made a number of improvements in map making. His reputation grew and in 1751 he was elected to the chair of mathematics at Göttingen University; three years later he was superintendent of the observatory, where one of his early important works was a meticulous observation of the libration of the moon. This was followed by his outstanding chart of the moon, published in 1775. His other published works included *Zodiacal Catalogue, Lunar Tables,* which were printed by the Board of Longitude in 1767, and the *Solar Tables* in 1770.

Mean Distance

The average of the least and greatest distances of a celestial body from its primary.

Mechanical Dial

One that is generally a type of equinoctial, with a radial index which is pointed towards the sun. The hour can be read from the tip of the index and the minutes will be seen on a smaller dial, indicated by a small hand which is geared to the hour-plate.

Mensula

A device that allows the drawing of a plan directly on a plane table whilst surveying. This is done by using the sight rules as drawing alignments. John Praetonius described one in about 1600.

Mercator, Gerardus (1512–94)

Flemish mathematician and cartographer born at Rupelmonde in Flanders. He studied first at Bois-le-Duc and then at Louvain. He met Gemma Frisius (*q.v.*) and became interested in geography and map making. Charles V became his patron and Mercator made a complete set of instruments for observation on the Emperor's campaigns;

sadly these were destroyed by fire and another set was ordered from him. Around this time he also produced his famous survey and map of Flanders, a copy of which is in the Musée Plantin at Antwerp. In 1538 his map of the world was published and in 1541 he produced his fine terrestrial globe. In 1551 his celestial globe appeared. In the middle of this amazing industry he found himself shifting towards Protestantism and he was arrested and duly prosecuted for heresy. Of the 42 people arrested with him, two were buried alive, two were burnt and one was beheaded but Mercator managed to escape. He worked with Ortelius (*q.v.*) to free the study of geography from the tyranny of those who still adhered to the teachings of Ptolemy (*q.v.*).

Meridian

May be both a terrestrial and a celestial great circle. The terrestrial meridian of any

Meridian solar gun, a toy for the rich collector. As the Sun rises the rays focus through a glass on to the touch-hole of the mini-cannon.

place on the earth is the line running through it and through the north and south poles. For each terrestrial meridian there is a related celestial meridian which passes through the north and south celestial poles and will also pass through the zenith and nadir of any place on the terrestrial meridian.

Meteor

The word is from the Greek meaning literally 'things in the air'. The ancient Greeks applied the name to many atmospheric phenomena – shooting stars, halos and rainbows. A meteor becomes incandescent when entering the earth's atmosphere, this light is caused by the heat friction on its surface. Most of the meteors are tiny, not more than a few millimetres across and as they enter the earth's atmosphere their speed may be as high as 50 km a second. The composition of these bodies may be either predominantly iron or stone and they can be as large as 100 tons. One of these monsters was brought to the United States by Robert Edwin Peary (1856–1920), the American explorer, who found it at Cape York, Greenland; known as the Anighito meteorite it weighed around 100 tons. A meteor that survives the entry of the earth's atmosphere and falls to ground is known as a meteorite. Our atmosphere is assaulted daily by probably up to 1,000,000 meteors and the waste material from their destruction in the atmosphere can add about 10 tons to the earth's surface.

In early times it was thought that meteors were generated in the air by inflammable gases. The odd fireball and star showers, shooting stars, had been noted but they were not attentively observed as their appearance drew forth feelings of dread and superstition, even forebodings of great malevolence. When found by primitive peoples because of their heavenly origin, they are regarded as objects of wonder and soon become wrapped in legends.

Multitudes of meteors are at large in outer space. Stand outside on a clear moonless night and patience should provide sightings of up to ten shooting stars in one hour. The month of August can often give a rich display of these brilliant heavenly fireworks. The smallest of them may be no more than the size of a coarse grain of sand and yet they still appear white and bright as they enter the atmosphere. Those that survive the furnace of the air and reach the earth become precious not only for veneration but also for scientists to analyse. Scholars apparently did not take them seriously and begin properly recording them until the end of the 18th century.

One of the most celebrated of the early arrivals came down in Phrygia and for many generations was worshipped there under the name of Cybele, the mother of the gods. A soothsayer told of how the possession of the meteorite would bring prosperity to the Romans and it was demanded by them that King Attalus should deliver it. When it arrived in great state in Rome a historian described its appearance as 'a black stone, in the figure of a cone, circular below and ending in an apex above'. It is likely that one of the holiest of Moslem relics could be a meteorite; built into a corner of the Kaaba is a black stone believed to have been given by Gabriel to Abraham. Although as mentioned above large numbers of small meteorites do survive and fall on the earth, the odds of being hit are extremely small, as is the likelihood of damage to property because the area of the earth which is covered with buildings is very small.

Meteorograph
An instrument that can give a continuous record of fluctuations in the humidity, pressure and temperature of the atmosphere. It is strictly a combination of a barograph, hydrograph and thermograph.

Meteorology
The science of the study of the earth's atmosphere and the conditions and changes in its weather. Probably the earliest mention is an observation by Aristotle (q.v.) in his book *Meteorologic*; but little was done in systematic observation or precise recording until the 16th century when Galileo (q.v.) and other Florentine scholars made a workable thermometer and in 1643 when Torricelli (q.v.) developed the principle of the barometer. The research of Boyle (q.v.) and Hooke (q.v.) advanced matters and Fahrenheit's thermometer gave more reliable readings. The Meteorological Office was established in London in 1854, founded as a department of the Board of Trade under Admiral Fitzroy. Six years later he began the first collections of weather reports by telegraph.

Meton (active 432 BC)
Early Greek astronomer associated with the introduction of the cycle known as the Metonic cycle. This is a period of 19 years during which there were 235 lunations. The calendar related days, months and years in a

way that did not accumulate discrepancies and was surprisingly accurate. Modern calculations have shown Meton to have been only slightly adrift.

Metronome

An instrument for directing the speed at which a musical composition should be played. The invention has been falsely ascribed to Johann Nepomuk Maelzel (1772–1838), who gave the date of his device as 1814. The earliest mention of such an instrument is 1696, when a description of a weighted pendulum of variable length appeared in a paper by Étienne Loulié. Harrison (*q.v.*), the famous maker of chronometers, in 1775 published a description of such a device.

Michell, John (1724–93)

English natural philosopher and geologist, educated at Queen's College, Cambridge. His research and inventions covered the torsion balance, which was later to be further developed by Coulomb (*q.v.*); magnetic observations and other geological and astronomical discoveries. His published works included *A Treatise of Artificial Magnets, in which is shown an easy and expeditious method of making them superior to the best natural ones* and *Conjectures concerning the Cause and Observations upon the Phenomena of Earthquakes.*

Micrometer

A device for accurately measuring very small angles or spaces.

Microphone

A device used in sound reproduction systems for converting sound waves into electrical energy which can then be reconverted into sound after transmission by wire or radio. It was primarily the invention of David Edward Hughes (*q.v.*) the Anglo-American electrician.

Microphotometer

A sensitive densitometer, an instrument for measuring the optical density of a material by recording the reflection of a beam of light directed on to the specimen.

Microscope

An optical instrument that enables the eye to see and accurately examine images of very small objects which would otherwise remain unobserved. It uses a lens or combination of lenses; modern microscopes can produce magnifications of up to 2,000 times.

Microscopes are principally of two types: the compound, which has two converging

19th century brass 'Culpeper' microscope, drawer containing accessories and with pyramid-shaped case

19th century brass compound monocular microscope, constructed on the principle of 'Joneses Most Improved', complete with test slides and fish-holder

Microscope slides by C. Collins

lenses or systems of lenses, these are the objective and the eyepiece; and the simple, which has a single lens. There are also the electron microscope, which uses electrons instead of light and electron lenses to give the enlarged image and the ultramicroscope, which is employed for studying colloids and has the light coming from the side so that the particles are seen as bright points on a dark background.

The history of the microscope starts about 100 BC when the philosopher Seneca used a globe of water to help him read and recorded that 'letters though small and indistinct are seen enlarged and more distinct'. Layard, the archaeologist, when excavating at Nimrud found a convex lens of rock crystal. The ancient gem-cutters were so perfect and detailed with their work that it is highly likely that they would have had some form of magnification to assist them

with this work and quite possibly they themselves would have made the lenses.

Descartes in his *Dioptrique*, published in 1637, describes a microscope with a concave mirror pointed towards the object used with a lens. The mirror lights up the object which is mounted on a point at the focus of the mirror. One of the first really successful grinders of high quality lenses was Anthony van Leeuwenhoek (*q.v.*); he also made them with a short focus. Another Dutchman, Musschenbroek (*q.v.*) from Leyden, was a successful maker of simple microscopes. Robert Hooke (*q.v.*), the Englishman, had the ability to shape the smallest of lenses which were constructed from glass globules made by fusing the ends of threads of spun glass.

The first compound microscope, the principle for which was probably discovered by Johann and Zacharias Janssen about

Late 19th century brass monocular microscope with coarse and fine focusing by Ross-Zentmayer, fitted with a set of six objectives, polarizer, analyser, compound condenser and bull's eye condenser

1590, was a system of a powerful biconvex with a still stronger biconcave lens. In 1646 Fontana described a compound with a positive instead of a negative eyepiece. About 1665 Robert Hooke (*q.v.*) produced a compound and 1668 the Italian Divini used pairs of plano-convex lenses. James Mann, the son of a Hertfordshire tailor, produced a compound with an adjustable tube allowing for variations in the enlarging powers of the system. It is interesting that, although he may have been a fine craftsman with his microscopes, he was not above sharp practice. The Spectacle Makers' Company in

Wooden barrelled microscope, stained vellum covered barrel, signed I. Marshall, *c.1715*

1695 fined him for selling spectacles ground on one side only. In 1672 Sir Isaac Newton (*q.v.*) proposed using concave mirrors. English makers were numerous and many of them top quality craftsmen including Edmund Culpeper (1660–1730), John Cuff

Mid 19th century brass compound monocular microscope, signed Powell & Lealand, 107, Euston Road, London, *including a parabolic illuminator, polarizing lens and opaque object prism*

(1708–72) (*see* Forgery, p. 245) and George Adams (*q.v.*).

Besides the microscopes mentioned above others for specialized use include binocular, demonstration, petrographical, inverted, multi-ocular, reflecting and 'museum'. The last is somewhat of an oddity, which carried a revolving cylinder with a large number of specimen apertures.

Milky Way *see* **Galaxies**

Mill, Hugh Robert (1861–c.1940)
Scottish meteorologist born at Thurso, he became a serious researcher into rainfall and in 1910 presented the rainfall details of the British Isles since 1677. He was also rainfall consultant to the Metropolitan Water Board.

Monge, Gaspard (1746–1818)
French mathematician, the inventor of descriptive geometry, born at Beaune. He was

educated first by the Oratorians in Beaune and then in their college at Lyons, where he was made a teacher at the age of 16. His published work included *Géométrie Descriptive*.

Moon

A satellite of the earth, some 238,800 miles away from it. The moon's diameter is about 2,163 miles and the earth's 7,918. The so-called 'phases' of the moon are actually the way we see the sun shining upon her surface during the course of her monthly rotations around the earth. The early Babylonians thought that the moon had a dark and a light side. Few objects can have gathered around themselves such a mass of superstition, symbolism and muddling myths as the sphere without an atmosphere that ceaselessly circles the earth.

In classical mythology the moon was called Hecate before she rose and after she had set; when a crescent she was Astarte, when high in the heavens Diana or Cynthia, she who hunts the clouds and as Phoebe was regarded as the sister to the sun. One legend has it that all the treasures that had been wasted on earth – misspent time and wealth, broken vows, fruitless tears, unfulfilled desires and intentions – were kept on the moon.

The first map of the moon was constructed by Galileo (*q.v.*); Tobias Mayer (*q.v.*) published another in 1775 and during the 19th century constantly improved maps were brought out by such people as Beer and Mädler, Schmidt and Neison. In 1903 Professor W. H. Pickering issued a complete photographic lunar atlas.

Moore, Francis (1657–c.1715)

English astrologer born at Bridgnorth in Shropshire. In 1700 he published *Vox Stellarum*, which was continued as *Old Moore's Almanac*.

Morse, Samuel Finley Breese (1791–1872)

American painter and inventor who was born at Charlestown, Massachusetts, son of Jedidiah Morse, a congregational minister and a writer on geography. In 1811 he journeyed to England to study painting under Benjamin West. He returned to America in 1815 and became first President of the National Academy of Design, New York. In 1832 he thought up the idea for a magnetic telegraph; in 1843 Congress granted him $30,000 for an experimental line between Baltimore and Washington. Actually Cooke and Wheatstone (*q.v.*) working independently in Britain just beat Morse into second place. These two had been working on it since 1837 and it was installed on the Great Western Railway for signalling and alarm systems. In many ways Morse's most internationally helpful invention was his Code – a system of long or short flashes or sound signals that could be used for communication by wireless, lamp or flags. He also worked at a marble cutting machine and experimented with telegraphy by submarine cable.

Mudge, Thomas (1715–94)

English clockmaker born at Exeter, the son of a clergyman who, when he saw his son's mechanical leaning, allowed him to be apprenticed to Graham. In 1758 he invented the lever movement but Mudge did not seem to realize the importance of his production and only made two or three examples, one of which was given by King George III to Queen Charlotte. He was awarded a substantial grant for his work with chronometers but Dr Maskelyne, then Astronomer Royal, did not agree with the award and argued with the Board of Longitude against it. Like Harrison's (*q.v.*) chronometer, Mudge's was sent off to sail the seas, this time to Newfoundland; the report on its performance from Admiral Campbell was excellent. Mudge tried to petition Parliament to get his money but even then enemies blocked the way.

Mural Arc

A type of telescope fixed to an arc which is graduated; the arc may be any shape from a quarter circle to a complete circle. The telescope and arc are fixed to a wall that is carefully aligned north and south so that as a star crosses the meridian, its exact position can be recorded.

Musschenbroek, Pieter van (1692–1761)

Dutch natural philosopher, born at Leiden where his father Johann Joosten was a maker of physical apparatus. He studied at the University in his native city. In 1723 he was appointed to the chair of mathematics at Utrecht and in 1732 to the chair of astronomy. He resisted the invitations of George II to join the newly founded establishment at Göttingen and went back to Leiden. His researches lay with capillary attraction, magnetism, the pyrometer and the Leyden jar. His most important book was *Elementa Physica*.

Napier, John (1550–1617)

Scottish mathematician, born at Merchiston near Edinburgh, one of the eighth generation of Napiers, the first of whom had acquired the Merchiston estate from James I of Scotland. John was educated at St Andrews University and on the Continent. About 1594 he started research which led towards his great discovery of logarithms, a task which was to take him the next 20 years. He published his findings in *Mirifici Logarithmorum Canonis Descriptio* in 1614. This was followed three years later by *Rabdologiae seu Numerationis per Virgulas libri duo* which was a description of an ingenious method of multiplying and dividing with 'bones' or 'rods'. By his work, particularly the *Canonis Descriptio*, John Napier had made a great breakthrough in the field of practical mathematics and by so doing placed himself alongside other searching minds of the period – Brahe (*q.v.*),

Copernicus (*q.v.*), Galileo (*q.v.*) and Kepler (*q.v.*).

It is of interest to consider the environment in which Napier lived – Scotland then in many places was a wild and lawless territory, unsettled, superstitious and credulous of secret powers. Like Kepler and other contemporaries, Napier believed in astrology and also had some faith in the ways of magic. A contract written in Napier's own hand exists between himself and a rough and tough Baron Robert Logan of Restalrig, in which Napier states he will undertake, '. . . to serche and sik out, and be al craft and ingyne that he dow, to tempt, trye, and find out' some hidden treasure that legend had it was hidden at the Baron's hideout at Fastcastle. For this Napier was to have one third of the treasure recovered.

The document is signed, 'Jhone Neper, Fear of Merchiston' and dated July 1594. However, in his scientific writings there is no hint of astrology or dark magic.

Napier's Bones
What amounted really to an early pocket calculator. A set would consist of graduated rods or loose cubes with a series of numbers divided by diagonal lines. The bones are set in a tray. Another version, however, has cylindrical bones that can be turned in their box.

Nasmyth, James (1808–90)
Scottish engineer born in Edinburgh, the youngest son of Alexander Nasmyth, the Scottish landscape painter. From an early age he was mechanically minded and in

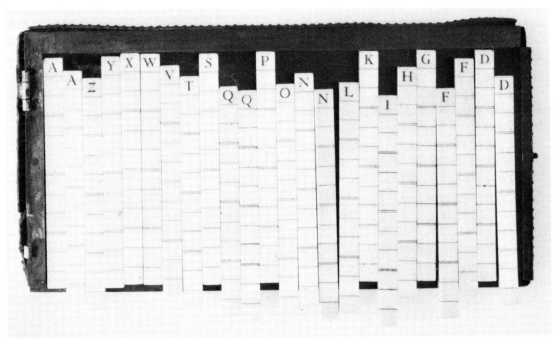

Set of twenty-four bones of square section for calculating, related to Napier's Bones, c.1700

1829 obtained a position at Henry Maudslay's (*q.v.*) works in London. In 1839 he invented the steam hammer with which his name is associated. His claim for the invention was disputed by the French manufacturer Schneider, who had actually copied the design from one of Nasmyth's notebooks and built an example at his Creusot works. In any case, it is likely that Nasmyth himself was actually beaten to the idea by James Watt. Nasmyth did, however, develop and improve machine tools like the nut-shaping machine, planing machine, steam pile-driver and other hydraulic machinery. In his retirement he spent much time observing the heavens, in particular the moon and later he published *The Moon considered as a Planet, a World and a Satellite.*

Nautical Almanac
A volume used by navigators and astronomers; it is published annually several years in advance. The first one was brought out in 1767 by the Royal Astronomical Society, who published it until 1834 when it was taken over by the Admiralty. Amongst the information it contains are tables and calculations of the tides and times of the sun's meridian.

Nautical Mile
Equal to 1,852 m (6076.103 ft), it was formerly called geographical mile and today is also termed international nautical mile.

Navicula Dial *see* **Regiomontanus Dial**

Navigation
The science or art of conducting a vessel across the seas. The early sailors of the western world, Phoenicians, Carthaginians, Greeks and Romans, who were sailing before the introduction of a compass, more or less had to keep within sight of land or perhaps on clear nights raise their eyes to the heavens and take some guidance from familiar stars. To the east the mariners of China and India noted that, during the monsoons, the winds blew with a steadiness of direction so that if they ran with the winds they could make long journeys in safety out of sight of land at one season and make the return trip when the winds were reversed.

It was the commercial sailors from Portugal during the early part of the 15th century who really began to make progress with the science of navigation, epitomized by the work of Prince Henry the Navigator (1394–1460), the son of John I of Portugal. For 50 years Prince Henry despatched his sailors on voyages of discovery which he financed and organized. The ships sailed down the west coast of Africa, to the Azores and the Cape Verde Islands; an observatory was built at Sagres and an early school of navigation was established there. John II, who came to the throne of Portugal in 1481, maintained the progress. He set up a committee of navigation, the tables of the sun's declination were worked out, an improved astrolabe replaced the cross-staff and in general a sound basis was laid for instruction for the navigators. In 1537 Pedro Nunez, cosmographer to the King, brought out a work on astronomy, charts and points regarding navigation; he saw errors in previous works and sought to correct them. Eight years later Pedro de Medina pub-

Ptolemy (*q.v.*) despite the work of Copernicus (*q.v.*). In 1556 Martin Cortes produced his *Arte de Navigar* with some illustrations of instruments, tables for the sun's declination, a year's reckoning of saint's days and details of charts.

In 1559 an Englishman, William Cunningham, brought out *Astronomical Glass*, which had some instructions on the making of charts. In 1581 Michael Coignet from Antwerp was publishing charts and within three years the Dutch had charts made up as atlases. Mercator (*q.v.*) with his maps and globes brought yet more light. In the final decade of the 16th century the impetus for captains of the seas continued with the publication of *The Seaman's Secrets* by John Davis; in this he proposed to give all that was necessary for navigators. Edward Wright of Cambridge brought out *Certain Errors in Navigation Detected and Corrected* and in 1637 Richard Norwood, himself a sailor, published *The Seaman's Practices*.

The sound basis for navigation came in the first half of the 18th century with the invention of the sextant and the chronometer. These two instruments brought a greater safety on the seas, which was added to by expanding surveys, improved charting and chart making and instructional volumes such as John Robertson's *Elements of Navigation*, which ran through six editions between 1755 and 1796, and Nevil Maskelyne's *British Mariner's Guide*. In 1795 the Admiralty Hydrographic Office was opened to make available an organization that has grown to assist the sailor.

Steel nocturnal, French, late 17th century

lished *Arte de navigar*, which was probably the first book devoted to the art of this subject, although he still leaned heavily on

Nebula

In astronomy, the name given to particular luminous cloudy patches in the heavens. If examined with a powerful telescope a nebula can be seen to be composed of clusters of small stars.

Negretti, Enrico Angelo Ludovico (1817–79)

Anglo-Italian optician born in Como, who settled in London in 1829. After training as a glass-blower he set up in business making thermometers. Later he went into partnership with Joseph Warren Zambra and built up a considerable reputation for producing high quality optical and scientific instruments.

Nephoscope

A device for calculating the speed of objects in the sky by noting their time of transit.

Newton, Sir Isaac (1642–1727)

English natural philosopher born at Woolsthorpe in Lincolnshire. His early education was at local village schools and continued at the grammar school at Grantham. Here at first he showed little promise and by his own confession was lazy until, attacked by another boy, he courageously defeated him and this must have released the stimulus which was to carry him forward. He reputedly made dials, kites, water clocks and windmills. In 1661 he was admitted to Trinity College, Cambridge. From here on Newton flourished. Sir David Brewster (*q.v.*) said of him,

To the highest powers of invention Newton added what so seldom accompanies them, the talent of simplifying and communicating his profoundest speculations.

S. Brodetsky said,

The three subjects to which the youthful Newton devoted the years immediately following his graduation, and in which he made such remarkable discoveries so early in his career, were an epitome of his life's work. Fluxions, gravitation and optics – these were the main concern of the whole of his scientific life.... No other man in the history of science presents a record of devoted service in the cause of science with such conspicuous genius and success.

Between 1665 and 1666 he discovered the binomial theorem. In 1668 he invented his reflecting telescope and throughout the rest of his life unleased a rich stream of advanced thought, research and invention: the law of gravity, the law of inverse squares, deduced from Kepler's third law, that could be applied to the motions of the moon, the use of a prism to split sunlight into a spectrum and the laws of motion. His reading interests encompassed theology, alchemy and ancient prophecies. He was a prodigious writer and the treatises and books he left included *Analysis per Equationes Numero Terminorum Infinitas* and *Philosophiae Naturalis Principia Mathematica*. Credit to his work and an understanding of the man was given by Sir David Brewster with his *Memoirs of the Life, Writings and Discoveries of Sir Isaac Newton* published in 1855.

Nivellator (or Nevellator)

A form of pocket line level that will allow

for readings up to a distance of 30 ft; it will also accurately determine gradients. Principally a tool used in road construction, drain and pipe laying, bricklaying and roof works.

Nova

A faint star that quite suddenly increases its brilliance by perhaps as much as 10,000 times in a very short period; this it may hold for a few days and then gradually the brightness fades, a process which can take months or even years. Tycho Brahe (*q.v.*) observed a very bright nova on 6 November 1572 in the constellation of Cassiopeia and in October 1604 Galileo (*q.v.*) and Kepler (*q.v.*) both saw one in Ophiuchus. The taking of photographs of the heavens has greatly assisted the spotting of novae.

Nuremberg Hours

In medieval times Nuremberg and towns close by had their own system of hours; these were reckoned as night hours from 1 onwards from sunset and daylight hours from 1 onwards from sunrise. This system produced the complication that at the summer solstice there would be 16 daylight and 8 night hours; for the winter solstice there would be 8 daylight and 16 night hours.

Nutation

A small periodic variation in the precession of the earth's axis which causes the earth's poles to oscillate about their mean position. It has a period of 18 years. The effect is caused by the moon's regular travel south and north of the plane of the orbit of the earth. In exact reckoning there are three nutations: lunar, monthly and solar.

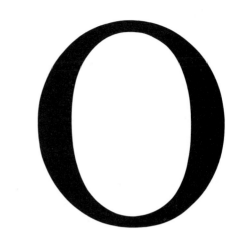

Object Glass

The main lens in a refracting telescope.

Objective

In optics, the lens or combination of lenses nearest the object.

Obliquity of the Ecliptic

In astronomy this is the angle between the plane of the earth's orbit and that of the celestial equator.

Occultation

In astronomy, the temporary hiding of one celestial body by another as it or they move in their orbit. To be accurate, the eclipse of the sun by the moon is an occultation of the sun by the moon.

Octant

An instrument for measuring angles; it is similar to a sextant, and a reflecting quadrant.

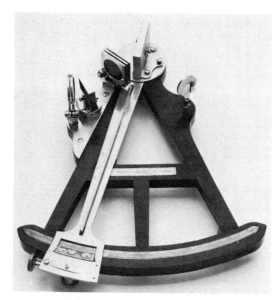

*Early 19th century English octant by
T. Hemsley & Son, Tower Hill, London, with
ebony frame and ivory scale and name plate*

Off-Centre Pendulum

One not centrally hung. The system was used in some of the wooden movement shelf clocks made in America by the likes of Eli Terry and Seth Thomas.

Ogee Clock

One in which the case has a reverse moulding curve. It is likely that the style started in Connecticut around the first quarter of the 19th century. The timepieces were generally shelf clocks but could be hung on the wall; they were spring or weight driven and had wood or brass movements of eight-day or 30 hour capacity.

Ohm, Georg Simon (1787–1854)

German physicist born at Erlangen and educated at the University there. He became professor of mathematics at the Jesuit College at Cologne in 1817, in the polytechnic school at Nuremberg in 1833 and professor of experimental physics at Munich University in 1852. His most important accomplishment was the discovery of the law describing the flow of an electric current through a conductor. In 1827 he formulated Ohm's law which is basic to the study of current electricity; this law states that the current flowing through a conductor is proportional to the voltage across it and inversely proportional to its resistance. From this, Ohm's name has been internationally incorporated in the terminology of electrical science in that an ohm has become the recognized unit of resistance. His best piece of written work was a pamphlet *Die galvanische Kette mathematisch bearbeitet*.

Ohmmeter

One that can be used to measure the insulation-resistance or other high electrical resistances.

Optical Illusion

A mirage, a phenomenon due to reflection and refraction of light during unusual atmospheric conditions. The commonest occurrences are in areas of calm in which there may be great heat or cold such as deserts or polar regions. The illusion may be seen during hot summer months in miniature over asphalt roads – the surface appears to be quivering in the distance ahead. A variation is the appearance of 'looming', an object may appear larger and nearer than it actually is. At sea objects that are still below the horizon may become visible from the bridge of an approaching ship. Instances of this are the 'Spectre of Brocken' and the 'Fata Morgana'.

Optical Telescope

One that is used to observe bodies by the light that they emit, as opposed to the radio telescope which is used to observe the radio frequency emissions. *See* Reflecting *and* Refracting Telescopes.

Optics

The science that deals with the phenomena of light and vision. It may be divided into three branches: physical, physiological and geometrical.

Orbit

In astronomy the path of any body, particularly a heavenly body, revolving around an attracting centre. The track is generally a well defined ellipse.

Orion

One of the brightest and most fascinating of the constellations, it contains three stars of fine quality – Rigel, Betelgeuse and Bellatrix – and 44 stars with magnitudes between 4 and 5.2. The whole constellation appears as a vast spiral nebula. Homer mentions Orion and it also appears in the Old Testament in *Amos* 5, v 8 and *Job* 9, v 9. Ptolemy (*q.v.*) catalogued 38 stars in the constellation, Tycho Brahe (*q.v.*), 42 and Hevelius (*q.v.*) 62.

Orrery

An astronomical instrument consisting of an apparatus which shows the principal movements of the solar system. The members are represented by different sized spheres and their movement in their orbits is controlled by arms and uprights moved by geared wheels.

The original instrument to bear the name orrery was made by John Rowley of London in 1713. This was not a new invention but a near copy of one that had been made at an earlier date by the leading horologist and instrument maker George Graham, assisted by his renowned uncle, Thomas Tompion (*q.v.*). Prince Eugene of Savoy had commissioned the orrery from Graham and Rowley had managed to make his copy from it whilst it was waiting transport to Europe. The name orrery was coined when the Rowley copy came into the possession of Charles Boyle, fourth Earl of Orrery.

It cannot be said that the orrery was suddenly invented, it was really a development from a planetarium or a similar device to show the movement of heavenly bodies. Although no examples of the very early machines of this nature exist there are records that point to experiments in the field. It is possible that the Chinese in the third millenium BC were operating water driven devices of this type. Cicero mentions a complicated globe, made by Archimedes (*q.v.*) in about the year 225 BC, that is supposed to have shown the movements of the sun moon and planets. Ptolemy (*q.v.*) in his book, *Almagest*, mentions a globe which he had made which would show the movements of the sun and moon. Following the work of Copernicus (*q.v.*) there was a considerable fashion for making these instruments, both designers and constructors vying with each other to produce the most complicated and ornate examples. They progressed from hand-operated to clock-

The original and historically important orrery made by John Rowley and named after his patron the 4th Earl of Orrery

work mechanisms that could accurately ape the relative movements of the heavenly bodies. Orreries as a whole can only produce the effect of movement; it would hardly be possible to make scaled representations of the comparative sizes of the sun, moon and planets or to reproduce the distances between each.

One of the most prolific of the 18th-century makers of orreries was Thomas Wright, who followed John Rowley. He was appointed as mathematical instrument maker to George II and must have been something of a publicist, for he placed an advertisement in a book entitled *Description and Uses of the Globes and the Orrery* which was published in 1745. This personal testimony ran:

The great encouragement Mr Wright has had for above seventeen years past in making large orrery's with the motions of all Planets and Satellites, and the true motion of Saturn's Ring: has made him so ready and perfect, that Gentlemen may depend on having them made reasonable and sound, not liable to be out of order.

As may be seen by one he made for Mr Watt's Academy in Tower Street.

Another for His Majesty's Palace at Kensington.

Another for the New Royal Academy at Portsmouth.

Another for his Grace the Duke of Argyle (late Lord Ile).

And several other large ones for the Noblemen and Gentlemen...

The first orreries employed flat discs to carry the spheres representing the sun, moon and planets. Benjamin Martin, in about 1764, broke away from this method and made an example in which the spheres were carried on arms which in their turn were attached to co-axial tubes which could be turned at different speeds.

Since at least the 14th century clocks had been made with various dials that would show the movement of the main heavenly bodies, but in the 19th century a firm signing their work Raingo Frères à Paris produced a number of what in fact were really very elaborate and expensive toys. These were clocks with small orreries mounted on the top. The examples would record and show the time in varying parts of the world, days of the month, movement of the earth, sun and moon and also the age of the moon and the timing of leap years.

Glass orrery sphere, stencilled with the polar star, Great Bear and other constellations enclosing a terrestrial globe

Ortelius, Abraham (1527–98)

Flemish geographer of German extraction who was probably born at Antwerp although his family came from Augsburg. With Mercator (*q.v.*) he was one of the greatest geographers of his time. He began as an engraver of maps and in 1547 he entered the Antwerp Guild of St Luke. In 1560 when travelling with Mercator to Trier, Lorraine and Poitiers he felt himself inspired by the example of the master to devote himself to scientific geography and at the suggestion of his friend to compile an atlas. The *Theatre of the World*, which was to establish his reputation, was issued on 20 May 1570 by Gilles Coppens de Diest at Antwerp. It contained some errors: the outline of South America is faulty and in Scotland the Grampians are shown as being between the Forth and the Clyde. In 1573 Ortelius brought out 17 supplementary maps under the title of *Additamentum Theatri Orbis Terrarum*.

Almost as a hobby he collected antiques, coins and medals and a catalogue of these was published in 1573. In 1575 he was appointed geographer to Philip II of Spain on the testimony of Arius Montanus, who vouched for Ortelius's orthodoxy as his family had been under suspicion of breaking away to Protestantism. In 1584 he brought out his *Nomenclator Ptolemaicus*, a series of maps illustrating ancient history, both sacred and secular. Practically his last work was an edition of the *Aurei sacculi imago, sive Germanorum veterum vita*.

Oscillation

An object is termed to be in oscillation when it is swinging backwards and forwards as, for example, the regular swinging of a pendulum. The verb oscillate means to vibrate or to swing to and fro.

Oscillator

A device for converting direct current into alternating current.

Oscillograph

An instrument for recording variations in an electric current, variations of an oscillating quantity.

Oughtred, William (1575–1660)

English mathematician born at Eton and educated there and at King's College, Cambridge. On entering holy orders he left the University in 1603 and went to the rectory of Aldbury, near Guildford. In 1628 he was appointed to instruct the son of the Earl of Arundel, teaching him especially mathematics. His fame rests on the fact that he invented the slide rule. Apparently he met his end when he collapsed with joy on hearing of the vote for the restoration of Charles II.

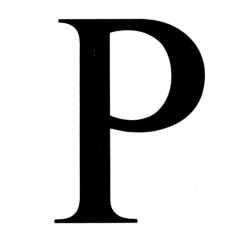

Pantograph

An instrument that can be used to make an accurate copy of a drawing to the same size or to enlarge or to reduce it. The normal form is an arrangement of flat bars of wood or metal, anchored in one place by a weight, and an upright on to which the pantograph is clipped. A stylus on one of the arms follows the lines of the original whilst a pen or pencil on another arm draws the copy. The size of reproduction is altered by changing the position of the drawing arm.

Brass pantograph by A. Abraham & Co., Liverpool, fitted with ivory wheels

Papin, Denis (1647–1714)

French physicist born at Blois on the Loire, he studied at the University of Angers and then moved to Paris. Here he met Christiaan Huygens (*q.v.*) and worked with him on his experiments with the air-pump. Later he crossed to England where he was welcomed by Robert Boyle (*q.v.*) whom he assisted with work in his laboratory and with his writings. His inventions included a double barrelled air-pump, a condensing pump and a steam digester for softening bones, which was described in a tract published in Paris as *La Manière d'amollir les os et de faire couire toutes sortes de viandes en fort peu de tems et à peu de frais, avec une description de la marmite, ses propriétés et ses usages.* This cauldron-like affair had a tightly fitting lid and used a safety valve for the first time. In 1679 he demonstrated that the boiling point of water depends on atmospheric pressure; this was illustrated by his pressure cooker for making 'bone broth'. He also discovered the principle of the siphon.

Pappus (active end of 3rd century BC)

Greek geometer who, in a period when mathematical science was retrograding, kept high standards and put down in his *Collections* much of his research. Although parts of this work are now lost enough remains to demonstrate his thinking. This work of Pappus was in eight volumes, the first of which has disappeared. From the rest we can see how he set out to arrange systematically a commentary on the work of his predecessors and earlier discoveries are enlarged upon. Amongst those scholars discussed are Apollonius, Aristarchus (*q.v.*), Eratosthenes (*q.v.*) and Euclid. Book VIII contains notes on mechanics, mechanical powers and the properties of the centre of gravity.

Parabolic Reflector

A concave mirror whose curve is part of a parabola which gives it the property of being able to reflect parallel rays to its focal point; it can be used as the object glass in a reflecting telescope. It can also be used for collecting and focussing an incoming beam of radiation. If used for light reflection it is more generally termed a parabolic mirror.

Parallax

In astronomy the apparent change in the

direction of a heavenly body when viewed from two different points. Owing to the rotation of the earth the parallax of a heavenly body alters, the maximum being when it is on the horizon. Annual parallax is calculated using the mean value of the earth's orbit as a base. During the 19th century considerable observation of parallax was undertaken. Sir David Gill (*q.v.*) mounted an expedition to Ascension Island to observe the phenomenon with Mars using a heliometer; later observations were made on Victoria in 1889, Iris in 1888 and Sappho in the same year. The heliometers used had been worked on by Gill and considerably improved from the earlier models. The results were discussed by Gill in the *Annals of the Cape Observatory*, Volumes VI and VII.

Parsons, Sir Charles Algernon (1854–1931)

English inventor and engineer, son of the third Earl of Rosse, he was educated mostly at St John's College, Cambridge. His primary invention was the marine steam turbine which bears his name. Others included a variation of the gramophone and non-skid tyres. The turbine was demonstrated by *Turbinis*, a launch, which startled Queen Victoria's Jubilee naval review at Spithead by attaining the then astonishing speed of 34½ knots. He published his father's work in 1926 under the heading *The Scientific Papers of William Parsons, 3rd Earl of Rosse*.

Pascal, Blaise (1623–62)

French mathematician and religious philos-opher, born at Clermont-Ferrand, on the Massif Central. His education appears to have been received largely from his father and from himself. A strange incident of his early years tells of his being bewitched and then freed from this thraldom by some old ceremonies. His undoubted mathematical ability showed itself at the age of 12 when he worked his way through Euclid, after which he wrote a book, which included a discussion of conic sections and drew praise from Descartes. He then followed up some of the experiments of Torricelli (*q.v.*). His other research encompassed determining the weight of air and from this evolving a means of measuring altitude by reading barometric pressure, working out his celebrated calculating tables, investigating the theory of probability and the differential calculus and the invention of a hydrostatic press.

In 1654 he joined the Jansenists and his thoughts were directed more and more out of the world towards eternity. His writings were numerous but amongst the most important are *Nouvelles Experiences sur le vide* (1647), *Lettres écrites à un provincial* (1656–7) and *Pensées* (brought out after his death). His contested feelings and passage through life are well summed up by Professor Clement C. J. Webb with this comment:

He was a man of science and he was also a saint, who had this indeed of the philosopher about him, that he was not able as some religious men of science have found themselves able, to keep his science and his religion in separate compartments of his life, but thought about their relationship to one another and felt the need of a unified view of the world.

Peate, John (1820–1903)

American telescope maker, bricklayer and Methodist minister who was born at Drumskelt, Ireland and who, when only seven, emigrated with his father to Quebec and then to America, eventually arriving in Buffalo, New York in 1836. Nothing is known of his early years or what education he had. When he was 16 he entered his father's trade as an apprentice bricklayer, but his work must have been somewhat intermittent as at this time he seems to have been attending Oberlin College. In 1851 he became a full-time minister, with great success; at one revival meeting he converted some 500 souls. In 1859 he sailed to Europe, visiting Ireland and England and embarking on a walking tour of the Continent and the Middle East. On his return to America he started to study astronomy and it is likely he made his first telescope to assist in his explorations. He had a natural feel for craftsmanship and is said to have made a 6 in refractor or a 6 in reflector or both; one of these was mounted at Chautauqua and then used in Peate's observatory at Greenville. After this he concentrated exclusively on reflectors, producing a 7 in for India, a 12 in for Harriman University, a 15 in for Allegheny College, followed by a 30½ in in 1891. A 22 in for Thiel College appeared never to have been used or even unpacked when found in 1935. His greatest effort was the 62 in reflector for a new Methodist university, American University, Washington. The difficulties behind the casting and grinding of this great mirror perplexed most American glassmakers. Eventually the Standard Glass Plate Company succeeded; the huge piece of glass weighed 2,500 pounds. After it was polished the mirror was silvered and Peate said,

It was silvered and tried on the heavens in the starless region under Corvus, and under the very imperfect management of the mirror on telescopic stars, the report was as good as could be expected.

Sadly this great project was never completed as Peate died whilst details were still being considered and finance arranged.

Pedometer

A meter for the rough measurement of distance travelled. The normal type is operated by an internal pendulum that responds to the movement of the owner; the pendulum is connected to a ratchet wheel which in its turn works a series of wheels which work the hands on a dial indicating the distance. The result can only be approximate as the step of the user will from one time or other vary quite considerably. Early examples from the beginning of the 18th century were sometimes operated by a cord attached to one of the feet of the wearer and then passed up to the pedometer which would be hanging from his belt; each step would give a pull to a small lever on the pedometer which in turn would operate a pawl which would cause a pointer on a dial to rotate. Johann Willebrand of Augsburg produced this type. Most of the pendulum type would resemble a pocket watch in appearance.

A hodometer is a similar kind of instrument for measuring the distance travelled by a wheeled vehicle; it is more accurate as the device is worked by the revolutions of the axle. A cyclometer is another type that is used with bicycles.

Pendulum

A body so mounted that it may swing freely under the influence of gravity; it is a device that has considerable value for experiments in physics. A 'simple' pendulum is a theoretical concept as it cannot actually be produced and seen to work; the description of it is a weight or mass suspended by a weightless thread. When Galileo (q.v.) timed the bronze lamp in Pisa by counting the oscillations against the beat of his pulse, he was really observing a close proximation to a simple pendulum, and he found the oscillations were isochronous, moving with a regular beat.

The earliest form of pendulum was the 'bob' and it was used with the verge escapement in clocks; this was developed by Christiaan Huygens (q.v.). The Continental suspension was generally by a fine cord as opposed to the rod favoured by the clock makers of Britain. A double bob pendulum was one that was spring-suspended with the rod carrying two bobs, a variation that came at the end of the 18th century. Other types of pendulum included the compensation, which could adjust to the effects of heat and cold; the conical, which rotated in a circle hanging from the apex of a cone, it was the brain-child of Jost Bodeker, Bishop of Osnabruck, invented in 1578, but it has not been greatly used; the Ellicott, introduced in 1752 which used the difference of expansion in brass and steel; also the gridiron which employed the same principle; the mercury, invented by George Graham, which employs a jar of mercury for the bob; there is also the wooden pendulum in which the rod is made of quality straight grained hardwood and if this is carefully selected and varnished it can be as accurate as any because the wood will not be liable to expand or contract with heat or cold.

Two specialist pendulums are the Blackburn, a contrivance for tracing curves – this is done by having a device within the bob for releasing small quantities of ink – and the ballistic pendulum, which was used for measuring the velocity of a bullet or for testing the power of gunpowder. The bob consisted of a mass of wood which swings outward when the bullet is fired into it.

Perigee

In astronomy it is the position of the moon's orbit when she is nearest the earth.

Perihelion

The point of the earth's orbit when it is nearest the sun; this term also embraces the same circumstance for a planetary body. *See* aphelion.

Perimeter

The line or curve enclosing a plane area.

Periscope

An instrument which enables the viewer to observe something which is not in a direct line of vision; this is achieved by a system of lenses and prisms.

Perkins, Jacob (1766–1840)

American inventor and physicist born at Newburyport, Massachusetts. He began his career as an apprentice to a goldsmith. His inventions included a technique for engraving bank-notes on steel plates. This was successful and in 1818 he travelled to England and went into partnership with the

Late 17th century German perpetual calendar and combined aide-mémoire *signed*
M. Mettlin fecit, *the covers carrying scales for the moon's age, sunshine in hours and minutes,*
moonrise and moonset, sunrise and sunset, saints' days, periods of day and night hours and
the signs of the zodiac

English engraver, Heath. He carried out many experiments with the compression of water and measured the results with another of his inventions, a piezometer which was particularly effective when very high pressures were involved.

Perpetual Motion

Like the lure of gold for the alchemist, *perpetuum mobile* has been the prize that in the past spurred on many serious physicists. Today no scientist would be tempted to waste his time searching for the unattainable – principles make such an achievement impossible. The idea of perpetual motion is that there could be a machine which once set in motion could go on doing useful work forever without drawing on an external source of energy; in fact it would have to be a machine which in itself would create energy. One idea was that there could be a clock that would go on rewinding itself until the parts of the machinery itself gave way. The energy of such a device would in fact defeat the restrictions of friction and keep going. There were also numerous wheels that had highly imaginative arrangements of curved spokes which allowed steel balls to tumble down and turn the wheel by the force of gravity. One famous wheel was that of the Marquis of Worcester who died in 1667. He gave the following account of his machine in his *Century of Inventions* (article 56):

To provide and make that all the Weights of the descending side of a Wheel shall be perpetually further from the Centre than those of the mounting side, and yet equal in number and heft to one side as the other. A most incredible thing, if not seen but tried before two Extraordinary Embassadors accompanying His Majesty, and the Duke of Richmond, and Duke of Hamilton, with most of the Court, attending him. The Wheel was 14 foot over, and 40 Weights of 50 pounds apiece. Sir William Balfore, then Lieutenant of the Tower, can justify it, with several others. They all saw that no sooner these great weights passed the Diameter-line of the lower side, but they hung a foot further from the Centre, nor no sooner passed the Diameter-line of the upper side but they hung a foot nearer. Be pleased to judge the consequence.

This trick of the overbalancing wheel to obtain perpetual motion seems to have been first dreamed up during the 13th century.

An architect, Wilars de Honecort, noted in a sketchbook:

Many a time have skilled workmen tried to contrive a wheel that shall turn of itself; here is a way to do it by means of an uneven number of mallets, or by quicksilver.

Leonardo de Vinci (*q.v.*) also had similar ideas.

Another of the wheel hopefuls was Johann Ernst Elias Bessler (1680–1745), known as Orffyraeus. He worked his way through a number of experiments that ended with a large wheel 12 ft in diameter and 1 ft 2 in wide; the construction was of a lightweight wooden framework covered with oilcloth that concealed the interior. The whole thing was mounted on an axle that gave the appearance of having no external moving agent. It received the approval of the Landgrave of Hesse-Cassel and apparently ran unaided for eight weeks in a locked and sealed room in his castle at Weissenstein. It seems that it so impressed the mathematician W. J. 'sGravesande that he wrote a letter to Sir Isaac Newton (*q.v.*) describing his examination of the Orffyraeus wheel and stating that it was indeed worthy of further study. He notes, however, that he had not been allowed to examine the interior of the wheel. The inventor strangely then seems to have destroyed the wheel stating that there was difficulty with a licence for it and that he was dismayed with the impudent curiosity of 'sGravesande. One can only guess as to the true motive power – an endless supply of small boys or terriers.

Yet another plausible dream of the perpetual motion seeker was the delightfully simple and apparently convincing water-wheel, so designed that it fed its own mill-stream. One naïve gentleman even showed a waterwheel pumping water into its own buckets. Hydrostatic everlasting motion is yet another beckoning finger for the dreamer. Denis Papin (*q.v.*), who himself dwelt on such matters as bone digesters, exposed one of these ideas in the *Philosophical Transactions* for 1685. This gadget consisted of a cup of water that at the base had a tube which was bent upwards and then into the top of the cup; the brave inventor had in mind that the larger amount of water in the cup would cause it to overbalance and so cause the smaller amount of water in the narrow tube to run into the top of the cup and to go on doing it for ever. Capillary attraction has been another favourite with the experimenters. The idea that an endless band could raise more water by capillary action on one side than the other and so mysteriously effect everlasting motion was considered. Sir William Congreve (1772–1828) had an inclined plane over pulleys, an endless band of sponge, another endless band of heavy jointed weights and placed the whole over water which would then be drawn into action.

A Bishop Wilkins thought about a lodestone and felt that it could be so placed as to pull a steel bullet up a reclined plane that had a hole near the top so that before it got to the lodestone it would fall through the hole and so return to the position from whence it started and so again ascend the incline and again drop through the hole and that would therefore be perpetual motion. The experimenter did not suggest exactly how his new endless supply of energy could be used profitably.

Petrology
The science of rocks which is concerned with the study of the history, structure and composition which make up the portions of the earth's crust that are accessible for examination. The observation of thin slices of rocks under a microscope is one of the most important of petrographic techniques.

Photo-Electric Cell
An instrument used for the measurement and detection of light. It has many applications: in an exposure meter for a camera, for example, or to activate certain types of intruder alarms.

Photometer
A device for comparing the intensities of sources of light.

Photomicrograph
A photograph of a highly magnified object taken through a microscope.

Photophone
An instrument for transmitting sounds by means of a beam of light. The feasibility of such an idea was voiced by Graham Bell (*q.v.*) in 1878 and described by him at a meeting of the American Association in 1880: one had been made that could send transmitted speech for 230 yards.

Photosphere
The visible white hot radiating surface of the sun in which sunspots appear.

19th century brass horizontal petrological microscope

Piazzi, Giuseppe (1746–1826)

Italian astronomer born at Ponte in the Valtellina. He entered the Theatine Order in 1764 and then took the chair in mathematics at the Academy of Palermo in 1780 and managed to get Prince Caramanico to erect an observatory there. In 1801 he discovered the first asteroid between Mars and Jupiter and called it Ceres. He also worked on corrections to the previous estimates of the aberration of light and published two catalogues of fixed stars. Later he was appointed director of the Naples Observatory.

Picard, Jean (1620–82)

French astronomer born at La Flèche. He became professor of astronomy in the College Royal de France in 1655 and was one of those picked out by Colbert to set up the Academy of Sciences. By 1667 he was applying the telescope to the measurement of angles and found a new system of astronomical observation with the pendulum. He set out his accomplishments in *Histoire Céleste*.

Pin-Gnomon Dial

The name can be applied to any dial that has a gnomon in the shape of a simple straight upright or horizontal thin rod.

Planck, Max Karl Ernst Ludwig (1858–1947)

German physicist born at Kiel, he became professor at the University there and later at the Physics Institute, Berlin. His work in thermodynamics and radiation culminated in his law of radiation which led to his discovery of the quantum theory of radiation. His hypothesis was that energy from an oscillating particle is emitted not continuously but in discrete quanta, packets of energy. He was awarded the 1918 Nobel Prize in Physics for his work in this field. His written work included *Prinzip der Erhaltung der Energie, Vorlesung über der Theorie der Wärmestrahlung, Das Weltbild der neuen Physik* and *Wege zur physikal Erkenntnis*. The Planck Institute in Munich is a clearing house for many of today's scientific advancements.

Planet

A non-luminous celestial body illuminated by reflected light from a star such as the sun. There are nine known major planets in the solar system: Earth, Mars, Mercury, Neptune, Jupiter, Pluto, Saturn, Uranus and Venus. In ancient astronomy a planet was one of seven heavenly bodies which were: the sun, moon, Mercury, Venus, Mars, Jupiter and Saturn. Since Copernicus (*q.v.*) the term can be applied to any opaque body moving in orbit around the sun.

Planetarium

An instrument that can simulate the apparent movement of the sun, moon and planets against a background of constellations; that is, an expansion of the orrery.

Planetary Nebula

Very faint hot star surrounded by an envelope of expanding gas.

Planimeter

An instrument for measuring the overall area of a figure. This is achieved by fixing the anchor end of two hinged rods into the drawing board whilst the end of the other rod traces the boundary line of the area to be measured.

Plücker, Julius (1801–68)

German mathematician and physicist born at Elberfeld and educated at Düsseldorf and the Universities of Bonn, Heidelberg and Berlin. He then went to Paris where he was influenced by the French school of geometers. His primary research was into analytical geometry during his earlier years, later he turned to physics. His discoveries ranged from the nature of gases and improvements with the spectroscope to work on the Geissler (q.v.) tube. His written work included *Analytisch-geometrische Entwickelungen, Theorie der algebraischen Curven* and *System der Geometrie des Raumes.*

Poincaré, Jules Henri (1854–1912)

French mathematician born at Nancy where he was educated at the polytechnic school. After a professorship at Caen he went to Paris University and was elected to the Academy of Sciences in 1887. He was one of the most brilliant mathematicians of the 19th century. His researches and discoveries covered many fields including the theory of functions, kinetic theory of gases, electro-optics, thermomechanics and celestial mechanics. His writings included *Théorie des fonctions fuchsiennes, Electricité et optique* and *Les méthodes nouvelles de la méchanique céleste.*

Polar Dial

One in which the plane of the hour-scale is parallel with the axis.

Polarimeter

An instrument for measuring the amount of polarization of light.

Pole-Star Recorder

Invented by Professor Pickering, it is a type of telescopic camera that can record the circle of the Pole Star. From the trace on the photographic plate it is possible to estimate the cloudiness of the atmosphere at night.

Polyhedral Dial

One with a number of surfaces bearing scales. It might be a small wooden cube, or a box with paper scales pasted on its faces, or it might be made of stone for exterior observations.

Portable Barometer

A type that can be placed in a case and have the siphon or cistern closed and the vertical mercury tube plugged for transport.

Potentiometer

A device for measuring direct current electromotive force or potential difference.

Precession

In astronomy, the oscillation of the earth's axis; this is caused by the slightly unequal gravitational attraction of the sun and moon. The precession of the equinoxes is the fact that they occur slightly earlier each year due to the westward movement of the equinoxes on the elliptic. This is caused by the precession of the earth's axis.

Priestley, Joseph (1733–1804)

English chemist and nonconformist minis-

Gilt-metal polyhedral sundial and clock, c.1650

tricity, in which he had earlier become interested. In 1767 he was appointed to the Mill Hill Chapel, Leeds where he switched his religious leanings from loose Arianism to Socinianism and produced pamphlets attacking the government's behaviour towards the American colonies. His interest in astronomy had grown and he was offered the post of astronomer with Captain Cook for his second expedition to the South Seas but his unpredictable character worried the members of the Board of Longitude and the appointment was not made. He then found employment with Lord Shelburne who took him on his travels around the Continent and in Paris Priestley met Lavoisier. Back in England in 1780, he parted company with his patron who allowed Priestley an annuity of £150 for life. In July 1791 the Constitutional Society of Birmingham had a dinner to celebrate the fall of the Bastille. Priestley was asked but would have little to do with it and expressed his sympathy with the revolutionists. Those holding the opposite view sacked and burnt the chapel and his house and all the contents including books, manuscripts and scientific instruments. His life's work was erased. Fortuitously he was left £10,000 by his brother-in-law and so pulled up what was left of his English roots and crossed the Atlantic to settle in Northumberland, Pennsylvania. His researches covered studies of acids, ammonia, air, carbon monoxide, carbon dioxide, oxygen and sulphur dioxide. His writings were prodigious and included six volumes on *Experiments and Observations on Different Kinds of Air, History and Present State of Electricity* and *History of Discoveries relating to Vision, Light and Colours.*

ter born at Fieldhead in Yorkshire. A somewhat wild and outspoken character, who, in many ways, was his own worst enemy. He began his career in trade and then entered a dissenting academy, after which he was dissenting minister at Needham Market and at Nantwich. Three years later, in 1761, he became a teacher in Warrington. He went to lectures in Liverpool given by Dr Matthew Turner on chemistry, and continued studies on elec-

Primary Mirror

The large concave mirror in a reflecting telescope that reflects the light from the object and produces the primary image.

Prism

In optics, a triangular sectioned piece of glass with rectangular sides that can be used for dispersing light into a spectrum or for reflecting and deviating light. It can be used in spectroscopes, binoculars and periscopes.

Proportional Dividers

These have legs with points at both ends and an adjustable joint. When open the distances between the points at the two ends are proportional to the lengths of the corresponding legs. The dividers can be used for enlarging or reducing plans and drawings

and will enable a high degree of accuracy to be obtained. It is possible that they were invented by Justus Byrgius towards the end of the 16th century.

Protractor

An instrument usually in the shape of a hemisphere, made of thin sheet metal or wood and carrying scales showing up to 180 degrees. It is used for laying down or reading angles.

Ptolemy, Claudius Ptolemaeus (active first half of the 2nd century AD)

The celebrated mathematician and astronomer was a native of Egypt and lived at Alexandria; it is not certain where he was born, different people ascribe it as being in Pelusium or Ptolemais Hermii. Although

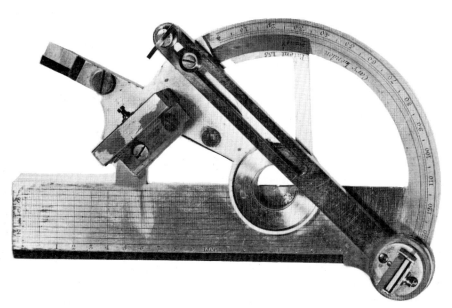

Douglas's patent reflecting protractor, c.1850

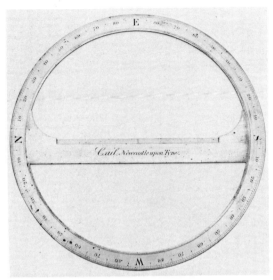

19th century brass circular protractor by Cail, Newcastle-upon-Tyne

finally proved wrong by Copernicus (*q.v.*), Ptolemy was really the greatest scientist of early centuries and his ideas were the basis for astronomical study until the Renaissance, a clear span of more than 1,200 years. The Ptolemaic system of the universe was quite simply that the earth was the centre of the universe and the heavenly bodies moved round it in circles. Earth, the stable element, occupied the lowest place, then water, the firmament and the ether; in concentric circular orbits revolved the moon – the nearest – and then Mercury, Venus, the sun, Mars, Jupiter, Saturn, then an immense 'crystalline' sphere into which the stars were fixed revolved round the edge of this self-contained universe. The system had snags which later astronomers tried to correct or alter to make it work. To the eight orbital spheres they added a ninth to explain the precession of the equinoxes, this

was followed by a tenth to make day and night. Then the sun sphere had to be placed eccentrically to take into account movement variation, a step towards thinking in terms of elliptical orbits. The puzzle became more and more intricate until Copernicus revealed a truth that may have been understood by him and other scholars but it still took nearly another 500 years to be universally accepted as correct.

Few exact facts are known about Ptolemy's early years. The Neoplatonic philosopher Olympiodorus related how Ptolemy devoted his life to astronomy and that he lived for 40 years, probably close to Alexandria, using the elevated terraces of the temple of Serapis at Canopus and raising pillars to record the results of his astronomical discoveries. It was also recorded that he made his first observation in AD 127 and the scholar himself said, 'We make our observations in the parallel of Alexandria.' Undoubtedly he followed the work of Hipparchus (*q.v.*) and in AD 150 updated his catalogue of stars. He must also have studied the records of the Chaldeans during the reign of their King, Merodach-Baladan, as he copied the details of a lunar eclipse on 19 March 721 BC which was observed in Babylon. The circular orbits that Ptolemy ascribed were a legacy from the much earlier astronomers who were convinced that the circle had such perfection that it alone was suitable for the movement of the heavenly bodies. In this field Ptolemy left his *Almagest* as a record.

He attempted to place geography on a scientific basis, outlining his thinking in *Geographike Syntaxis*. This was to have as much influence in the field of geography until the Renaissance as his *Almagest* was to have in the field of astronomy. Again he followed some of the thinking of Hipparchus who, three centuries before Ptolemy, had worked on ideas for making a map using latitude and longitude to mark the most important points on its surface. A reconstruction of the map of the world prepared by Ptolemy shows quite remarkable basic accuracy. Admittedly he had to a degree followed Marinus of Tyre, but most of the achievement was from his own talents. He also recorded the progress of Roman expeditions into the land of the Ethiopians and to Agisymba, possibly some part of the Sudan. The farther he got from Alexandria, the more Ptolemy had to rely on information that was sometimes of use but more often than not misleading. In northern Europe he managed to get the British Isles roughly in place but had Ireland too far north and Scotland twisted round so that its breadth was greater than its length. He supposed that the northern coast of Germany was the southern shore of the Northern Ocean; the Baltic was unknown as also was the Scandinavian peninsula. Turning to Asia, the Indian peninsula was flattened out. The placing of rivers is, at best, higgledy-piggledy.

In mathematics, some of what he did has been of lasting and constructive value, in particular his progress with plane and spherical trigonometry. He is also credited with writing on the musical scale and investigating optics.

Pyrheliometer

An instrument for determining the density of the sun's radiant energy.

Pyrometers

Instruments for measuring extremely high temperatures.

Pythagoras (6th century BC)

Greek philosopher, born at Samos or possibly one of the nearby islands, who is thought to have studied with Pherecydes. There is no doubt that he exerted considerable influence on the thinking of his time and later centuries. Heraclitus said of him, 'Of all men Pythagoras, the son of Mnesarchus, was the most assiduous inquirer.' He searched for answers, looked for reasons and travelled widely to bring himself into contact with surrounding peoples: those that he met could have included the Chaldeans, the Jews, the Phoenicians, the Egyptians, the Persians, the Brahmans and, moving westwards, the Druids of Gaul.

His mathematical research was the most important to him and it reflected his studies of architecture, astronomy and music. He settled at Crotona in the south of Italy, where he became the centre of a brotherhood of other awakening minds. Those who gathered round him explored not only scientific lines of thought but also ethics and the mystic doctrine of transmigration. The Pythagorean brotherhood spread over a wide area until its political associations led to a somewhat violent suppression. The meeting houses were looted and burnt, even in Crotona. Pythagoras withdrew to Metapontum, which is almost certainly where he died.

Of all the subjects that he and his followers investigated astronomy was the most important. They conceived of the earth as a globe out in space and revolving round a centre luminary with other planets; they did not, however, have the sun as this bright central feature but rather what they saw as some great fire, the hearth of the universe which they named Hestia. It was known that the moon was lit by reflected light and they thought that the sun also received reflected light probably from Hestia. To a degree they were ahead of their time in conceiving the heliocentric theory. Copernicus appreciated this and gave credit to the Pythagoreans that their thinking about the planetary movement of the earth gave him a first clue to the truth.

The well known Pythagoras' theorem about the right-angled triangle is, like other Pythagorean ideas, only associated with him by tradition.

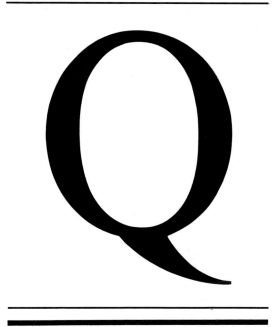

Pytheas of Massilia (4th century BC)

Greek navigator and astronomer, probably a contemporary of Alexander the Great. His work is lost and what scraps are known about his movements come from such as Polybius and are quoted by Strabo. It is said that he sailed from Massilia, today named Marseilles, westwards and then to the north and that he explored the coasts of Portugal, Spain, France and Britain. Strabo mentions in his Book IV, chapter I that Pytheas 'travelled all over it on foot'.

Pytheas was one of the first to make observations to determine latitudes; he also arrived at a theory for the tides with their connection to the moon and the relation of fluctuations to the lunar phases. Fragments of his thoughts were collected by Fuhr and published in Darmstadt in 1835 under the heading of *De Pythea massiliensi*.

Quadrant

An instrument formerly used by navigators to determine altitudes. Basically it consisted of a brass limb, a quarter of the circumference of a circle, and was graduated to one minute. The zero of measurement, or the reading of a vertical line on the limit, was carefully determined in the fixed instruments, whilst with instruments that were movable a plumb-line was used to mark the zero while an observation was being made. There were a number of varieties of quadrant and often these were known by the names of their inventors, such as Godfrey's, Gunter's (*q.v.*), Hadley's (*q.v.*), Sutton's etc, or they could be called by special names: mural quadrant, gunner's quadrant, surveyor's quadrant. The quadrant for navigational purposes has been completely replaced by the sextant.

18th century ebony and brass mounted quadrant by John Goater of Wapping, London

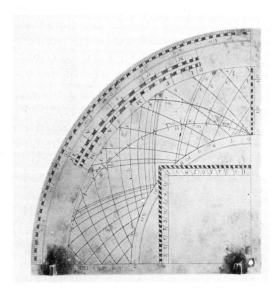

Brass Apian-type quadrant, early 18th century

Quadrature

In astronomy, the configuration when two celestial bodies, often the sun and moon or a planet, make a right angle with another body, usually the earth.

Quare, Daniel (c.1674–1724)

Leading English barometer and clockmaker, in business in London, first at St Martin's Le Grand and later in Exchange Alley. He was a Quaker and remained throughout his life a devout adherent of the Society of Friends. He even refused the important office of clockmaker to George I as he felt that his Quaker principles would preclude him from taking the oath of allegiance; the King rather sensibly waived the necessity of taking the oath and Quare took office. In 1680 he developed a repeating watch and some excellent timepieces that could run for a year on one winding and not only told the time but also marked the days, months and the difference between mean and solar time. He was Master of the Clockmakers' Company in 1708. A cutting from *The Daily Courant* for 2 May 1709 provides a rather touching comment on his nature:

Whereas Jos Wheeler, Brushmaker by trade, took away from Daniel Quare, Watchmaker in Exchange Alley, London, Money of Bills and Value; if any person can secure him, so as he may be brought to the said D. Quare, he shall have 10L reward; but if he surrenders himself with the said Money and Bills before the 7th of May, 1709, he shall be allow'd 20L. – Daniel Quare.

Quételet, Lambert Adolphe Jacques (1796–1874)

Belgian astronomer and meteorologist who was born and educated at Ghent. In 1819 he was appointed professor of mathematics at the Brussels Athenaeum; later he encouraged the building of the Royal Observatory in Brussels, was appointed its director in 1828 and succeeded by his son Ernest when he died. The elder Quételet's researches included shooting stars and other heavenly phenomena, magnetical meteorological observations and statistical investigations. His written works included *Sur l'homme et le développement de ses facultés, ou essai de physique sociale* and *Sur la Théorie des Probabilités*.

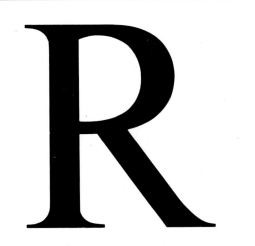

Radio Astronomy

The study of heavenly bodies by the detection and analysis of radio signals received on earth from radio sources far out in space; some of these sources may be undetectable by visible light.

Radio Carbon Dating

A technique developed and perfected by the American scientist Willard Frank Libby in 1947. It was primarily intended as a tool for the archaeologist and for use on objects with some carbon in their make-up. The method relies on the fact that the carbon in the air and tissues of living organisms have a few atoms of what is known as heavy carbon. Chemically the two carbons are similar, only varying in atomic weight – ordinary carbon is 12 and heavy carbon 14. Heavy carbon is radioactive which means that over a considerable period of time its atoms have gradually broken down by the emission of beta rays in a process that is termed radioactive decay. Research has shown that this decay and loss of radioactivity takes place in a regular manner, following the law of radioactivity decay, which is shown by the so-called half-life. With Carbon 14 the half-life is 5,600 years. The meaning of half-life is that, for example, if in the first place there were 1,000 atoms of heavy carbon in a substance, after a lapse of 5,600 years there will only be 500, after a further 5,600 years there will only be 250 atoms left as the rest will have decayed; thus, as time passes, the radioactivity will slowly be lost.

Libby found that by using an instrument like a very sensitive geiger counter he was able to measure the radioactivity of the heavy carbon in a sample of ancient wood and then compare the results with modern wood and a very old substance such as anthracite which, for all practical purposes, has lost all radioactivity. By applying the law of radioactive decay to his findings he could ascertain approximately when the particular tree that provided the sample had been felled. Radio carbon dating will not give an exact year, the results will follow what is called the standard deviation, which normally gives a datal area of about 300 years in which period the actual year will lie. For practical purposes the method can be used on specimens up to 40,000 years old. If they are older than this the deviation can become so large that the results will not be of much use.

Early tests with this technique were made with objects that could be proved against historically accurate chronological backgrounds. Wood fragments from tombs of Egyptian Pharaohs were used and the results came within 10 per cent of the known date. On Salisbury Plain, a date of 3,760 years for a piece of deer antler found close to one of the large stones at Stonehenge was obtained and small fragments from rope sandals found in Oregon produced a date of 7,000 years.

Radio Micrometer
An instrument for measuring or detecting radiant energy.

Radio Telescope
The instrument used in radio astronomy for picking up radio signals or for transmitting. There are two principal types of radio telescope. The first employs parabolic reflectors which are so mounted that they can be aimed at any part of the sky; incoming radiation can be reflected on to a small aerial at the focus point of the parabolic reflector. The second uses static radio interferometers, probably with two or more aerials each receiving signals from the same source and each coupled to the same re-

ceiver. In general the radio interferometer will be more accurate and have a superior ability to pick up a small source against a very dark background. The parabolic reflector, however, is more versatile because of its easily controlled manoeuvrability.

Raman, Sir Chandrasekhara Venkata (1888–1970)
Indian physicist who, after becoming professor of physics at Calcutta University, devoted his time to studying the problems of light and sound. In 1928 he found what has come to be called the RE – the Raman Effect. Basically, this is that when a monochromatic light passes through a transparent substance the emerging beam will consist of a mixture of light of the same wavelength as the incident beam, together with a small amount of scattered light of a longer wavelength.

Ramsden, Jesse (1735–1800)
English optician and instrument maker born near Halifax in Yorkshire. In 1758 he became apprenticed to a mathematical instrument maker and four years later was able to set up his own workshop. One of the first tasks he set himself was to improve the sextant. Additions he made included a small telescope to assist with sightings, a number of filters to reduce glare when looking in the sun, and the provision of a longer scale. He followed this with refinements for the theodolite and a mural quadrant. He invented a dividing machine for graduating his instruments. In 1777 he published details of this in *Description of an Engine for Dividing Mathematical Instruments*. Ramsden's most famous instrument must be a 5 ft

vertical circle which he made in 1789 for Giuseppe Piazzi (*q.v.*) to use at Palermo when he was preparing his catalogue of stars. His transit instruments were the first to be illuminated through the hollow axis, an idea that had been put to him by Professor Henry Ussher in Dublin.

Range Finder

This may also be termed a telemeter or position finder. It is an instrument designed to calculate the distance of an object with the minimum of trouble measuring a base. In 1891 the British Admiralty issued an advertisement for a range finder that would be suitable for placing on a ship. The main other requirement was that the device should be accurate to within 3 per cent at 3,000 yards. The resulting competition was won by the joint invention of Professor Bar of Glasgow University and Professor Stroud of the Yorkshire College.

Prior to 1908 the British Army had used the mekometer for ranging on to targets; this was basically a box sextant. In action two of them were used simultaneously from opposite ends of a fixed and measured base line. Its successor, the Marindin range finder, was invented by Captain A. H. Marindin of the Black Watch. The principle of the instrument was that of coincidence as with the Barr and Stroud naval model although a difference was that the right prism was made movable. The finder could be used vertically for horizontal targets or horizontally for upright objects.

Rankine, William John Macquorn (1820–72)

Scottish engineer and physicist born in

Edinburgh and studied at the University there. His principal work was on thermodynamics and he was the first writer to prepare a formal treatise on the subject. His other publications were *Manual of Applied Mechanics, Manual of the Steam Engine* and *Manual of Civil Engineering.*

Rayleigh, John William Strutt, 3rd Baron (1842–1919)

English physicist born in Longford Green in Essex, he entered Trinity College, Cambridge in 1861. His researches were conducted over a wide area, taking in such subjects as mathematical studies of electromagnetics and dynamics, determining the value of the Ohm, sound, colour and optics. He also did valuable work on wave motions, surface tension and in particular on the atomic weights of gases.

Réaumur, René Antoine Ferchault de (1683–1757)

French scientist born at La Rochelle who received his early education there. He later studied philosophy with the Jesuits at Poitiers and then civil law at Bourges under the eye of his uncle, canon of La Sainte Chapelle. His first intensive study was of geometry, but in 1710 he was diverted by a commission from the government to investigate and suggest improvements for manufacturing and also reviving forgotten industries. In 1731 he became interested in meteorology and invented the thermometer which bears his name. This took the freezing point of water as 0 degrees and its boiling point as 80 degrees.

Réaumur's scientific interests were so

wide that his colleagues called him the Pliny of the 18th century. He studied the strength of ropes, the tinning of iron, the forms of bird's nests, and the possibility of spiders being used to produce silk. He also produced *Mémoires pour servir à l'histoire des insectes,* in six volumes with 267 plates, published in Amsterdam from 1734 to 1742.

Reclus, Jean Jacques Elisée (1830–1905)

French geographer whose progress was marred because he had an overwhelming feeling for liberty; this led to his having to leave France in 1851 because of his outspoken support for republicanism. He travelled to America and the British Isles, later returning to France, where he again ran into trouble and was banished in 1872 for his activities during the Commune. He settled in Clarens in Switzerland and there produced his masterpiece, the 19 volumes of *Géographie Universelle.* This work was scientifically accurate and written with style; it earned for him the Gold Medal of the Paris Geographical Society in 1892.

Rectilinear Dial *see* Regiomontanus Dial

Red Giant

A huge star that emits red light. A star passes through a cooling phase once the usable hydrogen has been transformed into helium. This causes the outer shell to expand immensely and thus the temperature falls and the colour becomes red. The condition lasts until there are further reactions in the core, more energy is released and the shell is destroyed.

Red Shift

Possibly the result of the Doppler effect, caused by the recession of stars. It is a movement in the spectral lines of a stellar spectrum towards the red end when observed from earth; it can be taken to indicate that distant galaxies are moving away from our galaxy. This is evidence for the widely accepted theory of the expansion of the universe.

Reflecting Quadrant

Developed by John Hadley (*q.v.*) and the American Thomas Godfrey at the same time but independently. Because of double reflection, it only needed a scale over one eighth of a circle.

Reflecting Telescope

Also called a reflector, it is an instrument with which the initial image is formed by a concave mirror. James Gregory (*q.v.*) first thought of using concave mirrors for telescopes in 1663 but his scheme never materialized. Sir Isaac Newton (*q.v.*) turned to the use of mirrors and a reflected image, but he thought that it would not be possible to obtain lenses that would be free from chromatic aberration. The trouble with mirrors for telescopes was that until well into the 19th century all of them were made of metal alloys; it was possible to give them a high polish but depending on the constituents of the alloys the mirrors were prone to tarnishing and other forms of darkening. The particular alloy chosen by Newton was 'bell metal' which was composed largely of copper (usually six parts to two parts tin, with often some lead and zinc). After a period 'speculum metal' was employed; this

18th century 4¼in brass reflecting telescope signed Fran' Watkins, Charing Cross, London

was still primarily an alloy of copper and tin but the proportion of copper was less and the resulting mirror surface was clearer, harder and less likely to tarnish. The reflecting telescope was a great challenge to the craftsmen who ground the mirrors, as this work demanded great skill. One of the leading figures in the middle of the 18th century was James Short who specialized in the Gregorian reflector.

The arrangement of the primary and secondary mirrors and the eyepiece lenses can be noted in the photographs shown here. In the Gregorian, the secondary mirror is concave, whilst with the Cassegrain it is convex, in both cases the observation via the eyepiece is done through a hole in the centre of the primaries. Newton, in his reflector, used a different approach; he did not view through the primary but via a small flat mirror fixed by thin wires in the centre of the tube of the telescope and near to the open end; this small mirror was set slantwise so that it reflected the light rays straight into an eyepiece that was screwed into a hole in the side of the tube.

Sir William Herschel (*q.v.*) adapted Newton's idea with some of the huge telescopes he made and used. In his instruments the object glass or concave mirror was slightly tilted so that the light coming in from the object was reflected straight into the eyepiece which was fixed facing it in the side of the tube. The Herschel does have an advantage over other reflectors in that there is a saving of light as there is no second reflection. The disadvantage is that with small instruments the tilting of the mirror would bring an unacceptable distortion to the object being observed; but with instruments such as Herschel's 40 ft telescope with a 4 ft diameter this feature had only a slight effect. Lord Rosse (*q.v.*) constructed a reflector 56 ft in length and 6 ft in diameter at Birr Castle in Ireland. These examples were dwarfed by the immense telescope erected on Mount Palomar in America which had a 5.2 m (200 in) disk, a masterpiece of manufacture by the Corning Glass Works in New York. The instrument was designed by George Ellery Hale (*q.v.*) who had previously completed the large Hooker telescope on Mount Wilson in California. This monster which was finished in 1918, was a 100 in and the whole apparatus weighed 100 tonnes, most of which is carried by two mercury floats. Hale directed the design for the Palomar but sadly he died in 1938 and never knew the final creation of the giant which now carries his name. *See* Cassegrain *and* Gregorian Telescopes.

18th century brass reflecting telescope of Gregorian type, 5 in diameter barrel, fitted with geared altazimuth adjustment

Reflection of Light

When travelling through space, rays of light follow a straight path. On reaching another medium, part of the light energy is lost and the remainder is either bounced off as reflected light or penetrates and passes through the medium. The physical law is that the incident and reflected rays make equal angles with the normal to the reflecting surface at the point of incidence, and are coplanar with the normal.

Refracting Telescope

An instrument with which the resulting image is produced entirely by lenses, either convex or concave, set in line. It has been generally accepted that a Dutch spectacle maker, Hans Lippershey of Middelburg on the island of Walcheren, discovered the

19th century brass refracting telescope with mounted sighting tube and steadying bar

Brass refracting telescope by Nairne & Blunt, with steadying bar

principle, almost it seemed by chance, in 1608. Although the great Galileo (*q.v.*) quickly improved the performance of the refracting telescope, the instruments in early times produced only rather limited results, the main causes being the quality of the glass, spherical aberration and chromatic aberration. The telescope makers learned that by increasing the focal length of the instrument by reducing the curvature of the object glass both of these disturbing aberrations could be reduced. This process was highlighted in the 17th century by the 150 ft telescope constructed by Hevelius (*q.v.*) and the aerial or tubeless model by Christiaan Huygens (*q.v.*). The largest refracting telescope operating in the world today is the Yerkes at the Lake Geneva

Observatory about 80 miles north west of Chicago. It was erected between 1896 and 1897 and has an object glass about 40 in in diameter with a length of around 60 ft. This great scope was financed by the railroad millionaire, Charles Yerkes, who was persuaded to foot the bill of some $300,000 by the astronomer and designer George Ellery Hale (*q.v.*).

Refraction of Light

The change in direction of light rays which, when travelling at an oblique angle, encounter another medium that is transparent. The laws governing refraction are that the refracted and incident rays are coplanar with the normal to the refracting surface at the point of incidence. If the travel of the light rays is perpendicular to the refracting medium the refracted angle does not occur.

Refractometer

A device for measuring the refractive index of a substance.

Regiomontanus (1436–76)

German astronomer, his real name was Johann Müller, who was born at Königsberg in Franconia, the son of a miller. At first he called himself Joh. de Monteregio which later became altered to Regiomontanus. He studied at Vienna where he was a pupil of George Purbach and after a period became his associate working with the master on correcting the Alfonsine tables. Hindered by a lack of accurate translations of Ptolemy's (*q.v.*) treatises, in 1463 Regiomontanus went to Rome with Cardinal Bessarion to search for more authentic versions. Later he spent some time at the court of the King of Hungary but after an outbreak of war which diverted his royal patron the astronomer went to Nuremberg in 1471 and settled down to setting up an observatory where he undertook important work on observing eclipses and other astronomical phenomena. In 1472 Pope Sixtus IV summoned him to Rome to assist with the reform of the calendar.

He was materially assisted in publishing his work by Bernard Walther, a wealthy citizen of Nuremberg, who made available to him the means to start a book printing business. In 1475 Regiomontanus brought out *Tabulae Directionum* which was principally a work on astrology; another work, *Ephemerides*, contained valuable advice for determining longitude at sea. His posthumously published *Scripta* gave details of the many instruments Regiomontanus made with Walther's financial backing.

Regiomontanus Dial

This was a complicated vertical device sometimes found in conjunction with a nocturnal in the 16th and 17th centuries although those made in Italy were generally separate. It was also termed a universal rectilinear dial.

Reichenbach, Georg von (1772–1826)

German astronomical instrument maker who was born at Durlach in Baden. In 1804 he founded an instrument making business in Munich with Joseph Liebherr and Joseph Utzschneider and in 1809 with Joseph Fraunhofer (*q.v.*) and Utzschneider he opened an optical works at Benedictbeuern which was later moved to Munich. In 1814 he left both these concerns to set up another optical business with T. L. Ertel. His principal accomplishments included work on a dividing engine and developing Römer's (*q.v.*) ideas on meridian or transit circles for use in observatories. The Reichenbach version had one finely divided circle attached to one end of the horizontal axis, read by four verniers on an 'alidade circle', the unaltered position of which was tested by a spirit level. This device was accepted eagerly on the Continent; Bessel made one of the first in 1819.

Reinel, Pedro (active around 1504)

Portuguese hydrographer to whom is ascribed the innovation of placing a scale of latitude on charts; an example is preserved in the Bayerische Staatsbibliothek in Munich. This shows the technical hydrographic difficulty of integrating the scale of latitude, which is based on accurate astronomical measurement, into a chart which was originally drawn from estimated linear distance and with magnetic bearings uncorrected for variation. The chart shows the north east Atlantic and Mediterranean.

Resolving Power

In telescopes and microscopes, the ability to produce separate images of closely gathered objects. In a spectrometer, the ability to separate two adjacent peaks in a spectrum.

Rete or Spider

The term for the movable openwork front plate of a plane astrolabe or planispheric astrolabe. It normally carries the zodiac which may have pointers indicating fixed stars.

Rheostat

An electrical resistor which can be used to regulate current. It can be continuously varied.

Rhumb Line

The course of a ship that holds to a fixed compass direction; it is shown on a map as a line crossing all meridians at the same angle. It is also known as the 'loxodromic curve'.

Right Ascension

The celestial equivalent of terrestrial longitude; the angular distance of a celestial body or point on the celestial sphere, measured eastward from the vernal equinox along the celestial equator to the hour circle of the body or point. It is expressed in hours or degrees.

Ring Dial

One of the simplest forms of dial. Basically, it consists of a single ring with a sliding collar in which there is a small hole; this

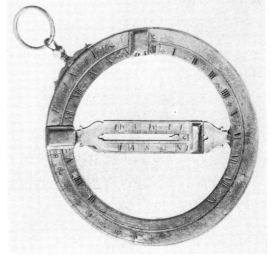

Ring dial signed Richard Abbott Fecit, *1668*

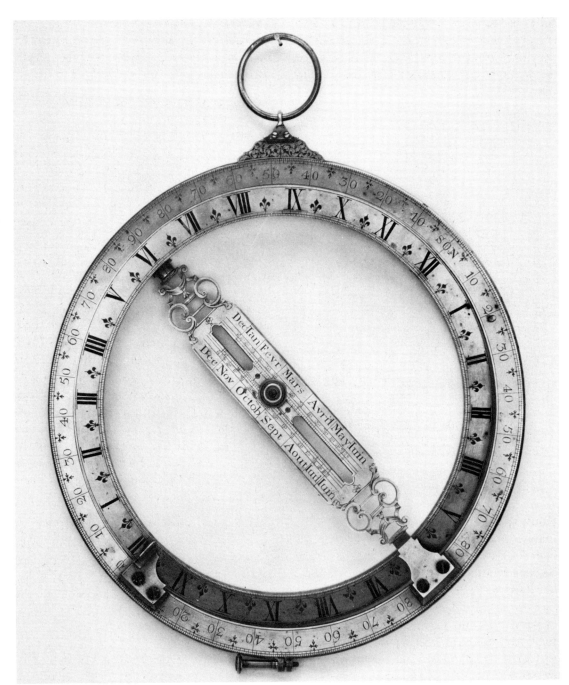

Universal ring dial by J. Rowley, 18th century

collar is set to the solar declination. When the ring dial is suspended and turned towards the sun the rays of light pass through the hole and hit the hour-scale marked on the ring.

Universal ring dial by Troughton and Simms, London, with silvered scales

Römer, Ole (1644–1702)

Danish astronomer who was born at Aarhus in Jutland. In 1671 he went to Paris where he was appointed tutor to the Dauphin. In 1681 he returned to Denmark and became director of the Copenhagen Observatory. His achievements included the invention of the astronomical instrument known as the meridian circle. He was also the first man to discover that light possessed a finite velocity, which he calculated from observing an eclipse of the first moon of Jupiter. His celestial observations were published by Horrebau in 1735 under the title *Basis Astronomiae.*

Röntgen, Wilhelm Konrad von (1845–1923)

German physicist, born at Lennep. He studied first in Holland and then went to Zurich to take his doctorate. He was for a time an assistant to Kundt at Würzburg. In 1895 whilst at Würzburg he discovered Röntgen rays. When he was experimenting with a highly exhausted vacuum tube on the conduction of electricity through gases, he noticed that a paper screen covered with barium platinocyanide which was lying nearby, became fluorescent under the action of some radiation emitted from the tube, which was at the time enclosed in a black cardboard box. Making further tests he found that this radiation had the power to pass through various substances which would be opaque to ordinary light. This was the beginning of X-rays. Röntgen also researched into elasticity, heat conduction by crystals, electromagnetic rotation of polarized light and the absorption of heat rays by different gases.

Rojas Dial

A vertical universal dial which was probably developed by Juan de Rojas, a talented follower of Gemma Frisius (*q.v.*). It is also called a Geminus and may be found on the reverse of some nocturnals.

Roman Strike

A device that was probably developed by Joseph Knibb and his brother towards the end of the 17th century. The purpose was to conserve the power required by spring driven clocks to drive the striking train. The Roman strike needed only 30 blows to give the run of 12 hours, as opposed to the 78 strokes required by the normal system. This was achieved by striking the hours on two different toned bells – one stroke on one bell indicating the I, II and III of the Roman numerals, one stroke on the other bell denoting the V and two strokes the X. Unless the face is clearly marked with Roman numerals the system could be confusing, the more so as at that time IV was sometimes marked as IIII.

Roscoe, Sir Henry Enfield (1833–1915)

English chemist born in London who studied first at Liverpool High School, then at University College, London, later going to Heidelberg to work with R.W. Bunsen (*q.v.*) with whom he collaborated on the development of comparative photochemistry. His published works included *Lectures on Spectrum Analysis, A New View of Dalton's Atomic Theory* and, in conjunction with Carl Schorlemmer, *Treatise on Chemistry.*

Rosse, Earl of, William Parsons (1800–67)

Irish astronomer and telescope constructor, born at York and studied at Trinity College in Dublin. Until his father's death he was known as Lord Oxmantown. From Trinity he went to Magdalen, Oxford, and in the same year (1821) he was returned as Member of Parliament for King's County in Ireland, a seat which he resigned in 1834; in 1845 he was made Irish representative peer.

His interest in astronomy and telescope making started in 1827; in 1839 he erected a telescope of 3 ft aperture at his seat at Birr Castle, Parsonstown and by February 1845 his 6 ft reflector was completed. But it had to stand to one side, as the famine and the unsettled state of the country demanded his attention as lieutenant of King's County; it was about three years before his lordship started to gaze out into the far reaches of the heavens.

The first maker of large scale reflectors had been William Herschel (*q.v.*) but he had never published anything about his methods for casting and polishing specula and existing records hinted that he had not been very successful with sizes over 18 in in diameter. Lord Rosse, therefore, had little to assist him with the task of producing his large scale reflectors. The speculum alloy he employed was composed of four parts copper and one part tin, this combination yielded a brilliant result with a high resistance to tarnish. Brewster's (*q.v.*) *Edinburgh Journal of Science* for 1828 carried a description of the machine Lord Rosse used for polishing speculum. It imitated the motions made in polishing by hand whilst the speculum revolved slowly; by shifting two eccentric pins the course of the polisher could be changed from a straight line to an ellipse of very small eccentricity, thus a true parabolic figure could be obtained. The great scope had a tube with a 7 ft diameter and a focal length of 54 ft; the whole was supported between two stout brick walls. The size and method of suspension did, however, constrict its use. It could not be pointed to every part of the heavens, it was only possible to move it a short distance from the meridian and very little to the north of the zenith. The Irish weather was another important factor. Birr was in the same area as the famous Bog of Allen and for much of the time it was either raining or about to rain. Apparently when the Astronomer Royal visited the site he found it 'absolutely repulsive'. Nevertheless, the 'large one', with its immense optical power, enabled many valuable observations of nebulae to be taken. On the demise of the 3rd Earl, his eldest son Lawrence, 4th Earl, continued his father's work with notable observations on the moon and studies of the radiation of the moon. A later member of the family published *Observations of Nebulae and Clusters of Stars made with the 6-foot and 3-foot Reflectors at Birr Castle from 1848 to 1878*; this was in the *Scientific Transcriptions* of the Royal Dublin Society.

Rotameter

An instrument for measuring the speed of flow in liquids. It consists of a small float, suspended by the fluid in a vertical calibrated tube. The height of the float indicates the rate of flow of the liquid.

Rule

A straight measuring instrument that can be made of wood, bone, ivory, metal or similar material. It is generally marked to show inches and fractions of inches or centimetres. A *pied de roy* is a folding foot rule.

Rutherford, Lord Ernest (1871–1937)

New Zealand physicist, born at Nelson and educated at New Zealand University and later at Cambridge, where he worked in the Cavendish Laboratory under Sir J.J. Thomson (*q.v.*). In 1898 he went to McGill University, Montreal, as Macdonald professor of physics, remaining there until 1907 when he returned to England to become Langworthy professor of physics at Manchester University and in 1919 Cavendish professor and director of the Cavendish Laboratory at Cambridge. His researches included electric waves, uranium rays (discovered by Becquerel), the discovery that radioactivity was an atomic phenomenon (in collaboration with Soddy), Röntgen rays and the ultimate constitution of matter; in the early 1920s he was the first to split the atom. His written work included *Radioactivity, Radioactive Transformations,* and *Radioactive Substances and their Temperatures.*

Safety Lamp *see* **Davy, Sir Humphrey**

Sagittarius

A large conspicuous constellation in the

southern hemisphere which lies between Scorpius and Capricornus on the elliptic and is crossed by the Milky Way. An ancient zodiacal constellation, the ninth in order, it represents a centaur drawing a bow. It was mentioned by Eudoxus (*q.v.*) in the 4th century BC and Aratus in the 3rd century BC. It was catalogued by Ptolemy (*q.v.*) as having 31 stars, Tycho Brahe (*q.v.*) 14 and Hevelius (*q.v.*) 22.

Sandglass (also Sandclock, Hour Glass)

An early device for measuring time. It consisted of two glass chambers linked by a narrow channel; the device contained a quantity of sand that would take a specific time to run through from one chamber to the other. The supporting frames for holding the glass chambers could be either of wood or metal. The exact origin is not known but it was probably first used in the 14th century. During the 15th and 16th centuries examples with four double chambers were produced for measuring the quarters of an hour.

Early 18th century brass mounted sandclock with 15 minute glass and two separately blown bulbs joined with wax cord

Satellite

The name given to a natural or artificial body orbiting a planet. The earth has one satellite, the moon, Mars two and Jupiter has eight. The name is derived from a Latin word signifying an attendant, for the bodies always move in close proximity to their respective primaries – the planets which they accompany.

Saturn

The most remote of the planets known to

Late 18th century French sandclock with half and full hour glasses

the ancients, modern measurement gives its distance from earth as 1,500 million km. Observations from the Voyager spacecraft have demonstrated that Saturn's rings are more complex than had been thought by earthbound observers. The planet rotates about every ten hours and takes around 29½ years to go round the sun. Her largest satellite is Titan. The early astrologers made much of the mystery and superstition that could be attached to this heavenly body; according to some of them Saturn typified lead and carried fearful consequences for those connected with it.

Observation of this body has attracted many of the famous in history. Galileo (q.v.) noted peculiarities in the planet's appearance in 1610; Huygens (q.v.) in 1655 was probably the first to offer a possible solution to the purpose of the rings. Cassini (q.v.) in 1675 observed the double nature of the rings and Bond and Dawes, acting independently in 1850, discovered the innermost 'gauze' ring.

Scale

An instrument for weighing; also, a sequence of regular marks that can be used for reference when taking measurements, a measuring instrument, the ratio between the size of something real and that of a model or drawing of it, or an established standard.

Scaphe Dial

One that is in the shape of a hemisphere and has the hour-lines engraved or painted on the inside of the hemisphere.

Schiaparelli, Giovanni Virginio (1835–1910)

Italian astronomer and senator of the Kingdom of Italy, who was born at Savigliano in Piedmont. He was first educated in Turin and then studied under Encke in Berlin. In 1859 he was made assistant observer at Pulkowa, moving in 1860 to a similar position at Brera, Milan, rising to the directorship in 1862, a position that he held until 1900. Schiaparelli was a patient and talented observer, his work over more than 30 years covered much ground. He showed that meteors or shooting stars traverse space in cometary orbits. His particular studies included double stars, the surface of Mars and its so-called canals, Mercury and Venus. When he retired he turned to the study of the astronomy of the Hebrews and the Babylonians. His written work included *L'Astronomia nell' antico Testamento, Note e reflessioni sulla teoria astronomica delle stelle cadenti* and *I precursori di Copernico nell' antichita.*

Schmidt, Bernard (1879–1935)

German maker of mirrors and lenses for amateur instruments who joined the team at Hamburg Observatory and there developed his telescope which was, in fact, a specialized camera named after him. It has become the standard instrument for photographing wide portions of the heavens. The largest Schmidt camera is at Tautenberg in East Germany and has a 2 m aperture.

Schmidt, Johann Friedrich Julius (1825–84)

German astronomer born at Eutin, he became director of the Athens Observatory.

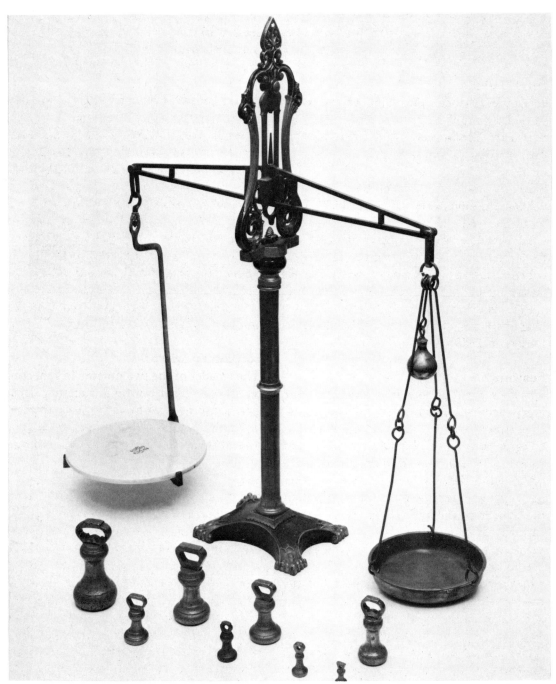

19th century beam scales, bearing original Avery trade label

He made a special study of the moon and discovered the new star in Cygnus, Nova Cygni.

Scintillation
In astronomy, the twinkling of the stars caused by rapid changes in the density of the earth's atmosphere, which produces an uneven refraction of starlight.

Scoresby, William (1789–1857)
English explorer, born near Whitby in Yorkshire. He made voyages in his father's whaler when a boy, later becoming mate on the *Resolution* in 1806. He made a valuable contribution to navigation by preparing an accurate chart of the east coast of Greenland which completely changed the then known characteristics of the area. He took Holy Orders in 1825.

Seasons
The four equal periods into which the year is divided by the solstices and equinoxes; they result from the apparent movement of the sun crossing the Equator from north to south and south to north during the earth's orbit round the sun. The earth as a planet has seasons independent of the distribution of land and water. As the earth moves through its elliptical orbit the changes take place. At the aphelion the northern hemisphere is canted toward the sun which then shines over the whole arctic circle and the northern hemisphere experiences summer whilst the southern hemisphere has winter, six months later the position is reversed. The nature of the earth's surface, the areas of land and water and the atmosphere can considerably modify the seasonal changes

and expected conditions. Changes of the seasons can often be marked by turbulent weather states.

Secchi, Angelo (1818–78)
Italian astronomer, born at Reggio in Lombardy, who entered the Society of Jesus at an early age. In 1849 he was appointed director of the College Romano Observatory which was rebuilt in 1853. During a life dedicated to the study of physical astronomy he concentrated on spectrum analysis and the physical constitutions of the sun, moon and planets. In 1863 he announced his system for classifying the spectra of stars and in 1867 his discovery of carbon in stars. His best known publication is *Catalogo delle stelle di cui si è determinato lo spettro luminoso,* published in Paris in 1867.

Secondary Mirror
The smaller of the two mirrors in reflecting telescopes. This mirror's function is to refocus the observed image and to improve the visual quality. The collection of light from the object is done by the primary mirror.

Sector
A measuring instrument with two graduated legs hinged at one end. The marked scales can include linear measurements, chords, sines or tangents and some sectors also carry degrees for measuring angles. A sector can also be an instrument whose limb embraces only a portion of a circle, which is used for measuring angles too great for the compass of a micrometer. When it is used for measuring zenith distances of stars, it is called a zenith sector.

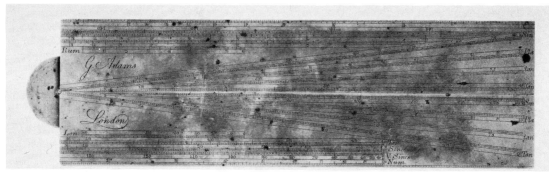

Brass calculating rule or sector by George Adams, London, 18th century

Finch brass sector, signed I:Finch fecit, *with hinged arms engraved with sets of lines of numbers, chords, sines, tangents and secants; English, mid-18th century*

Seeliger, Hugo von (1849–1924)

German astronomer born at Biala. He worked first at the Bonn Observatory and then in 1882 he was appointed director of the Munich Observatory and professor of astronomy at Munich University. His principal work was concerned with the distribution of stars; he also developed a theory on the meteoric constitution of the rings of Saturn.

Seismograph

An instrument for automatically recording movements of the earth's crust caused by earthquakes or tremors. The record from such an instrument is called a seismogram. The intensity of an earthquake is expressed on the Richter scale which registers from 0 to over 8; it takes its name from the American seismologist, Charles Richter.

Selenography

The division of astronomy which concentrates on the description and mapping of the moon. One of the most celebrated followers of this subject was John Russell (1745–1806) who was born at Guildford and who studied portrait painting with Francis Cotes RA. Russell was a man of strong religious convictions which reached every aspect of his life. In one of his diaries he made the inscription: 'John Russell converted September 30, 1764, aetat. 19, at about half-an-hour after seven in the evening.' He did in fact become a quite outstanding portrait painter, working in both oils and pastel, and his painting of ladies was especially remarked on. He was elected to the Royal Academy in 1788 and this was followed by an appointment as Painter to the King and Prince of Wales in 1790. Russell was a great friend of Sir William Herschel (q.v.) and it was from this scholar that he acquired his great love for astronomy; Russell was also a gifted mathematician. He invented a complicated piece of mechanism called the selenographia which was intended to exhibit the phenomena of the moon. A pamphlet explaining the machine was published and Russell prepared a great map of the surface of the moon, from which he engraved a series of plates to form a globe showing the 7/12ths of the moon's surface which is visible from the earth. The points of reference he took for the map were obtained from systematic observations using a telescope fitted with a micrometer which could measure the size of objects and details on the surface of the moon. The selenographia even had a support which allowed it to rock to simulate the libration of the moon.

Sextant

An instrument for measuring angles on the celestial sphere. It was invented by John Hadley (q.v.) in 1730 and afterwards improved by him; it enabled the navigator to take the angle between two objects. As it could be held by hand, it was possible to use it on the deck of a ship despite any prevailing motion. In use one object would be reflected on the other by means of a mirror; at the moment of coincidence the instrument would be clamped and the angle between the two objects read off from a scale. Shades of coloured glass were fixed to reduce the brightness of the object viewed directly or help to cut off the glaring brightness of the sun. Tycho Brahe (q.v.) made a number of simple sextant-like instruments which had two sights, one on a fixed and the other on a movable radius, so that the observer pointed to the two objects whose angular distance he was measuring.

The astrolabe and cross-staff had both proved themselves difficult devices with which to take altitudes and the underlying idea for an improved device seems to have entered independently into several different minds. Robert Hooke (q.v.) modelled two reflecting instruments, one of which, described in his *Posthumous Works*, had a single mirror to reflect the light from one object into a telescope which pointed directly at the other while the second used two single reflections. Its working was described in Hooke's *Animadversions to the Machina Coelestis of Hevelius* published in 1674. Sir Isaac Newton (q.v.) also thought about the problem but nothing came to light until 1742 when a description in his own handwriting of an instrument he had worked on was discovered amongst Halley's (*q.v.*) papers. This was to be a sector of brass, the arc of which, though only equal to ⅛th part of a circle, would be divided into 90 degrees. A telescope would be fixed along a radius of the sector with the object-glass close to the centre. Outside this a plane mirror set at an inclination of 45 degrees to the axis of the telescope would catch half the light which would have fallen on the object-glass. One object would be observed through the telescope whilst a movable radius, holding a second mirror close to the first, would be moved until the second object, by double reflection, was seen in the telescope to coincide with the first. Thomas Godfrey, a glazier from Philadelphia, produced a similar design to Hadley's working without any knowledge of what the Englishman was doing. Other makers included Jesse Ramsden (*q.v.*) and John M. Eckling, the latter sometimes producing pocket sex-

Brass double frame sextant by Troughton of London, 1815/16

tants fashioned in brass with a silver arc with a radius of around 4½ in.

A variation is the box sextant, a simplified device that is mostly used by surveyors.

Shadow
A dark toned image cast on a surface by the interception of light rays by an opaque body. Radiant energy travels in straight lines in the same medium; certain substances are opaque to one form or another of the waves. Thus on the lee side of the opaque objects there will be no manifestation of radiant energy and shadows will be formed.

Shadow Photometer
A device for measuring the strength of shadows by comparing intensities.

Sheepshanks, Rev Richard (1794–1855)
English astronomer who was born in Leeds and, although he took Holy Orders in 1824,

Brass sextant signed Ramsden, London

John Russell's selenographia, 1797

devoted himself to astronomy. He worked on Airy's (*q.v.*) pendulum experiments in Cornwall in 1828. His research on longitude enabled him to determine the longitudes of a number of places including Antwerp, Liverpool and Valentia using chronometric observations. He also helped to restore the standard weights and measures destroyed by fire in 1834. For a period he was secretary of the Royal Astronomical Society and personally he got together a fine collection of astronomical instruments.

Shepherd's Dial *see* Cylinder Dial

Sidereal

Of, or pertaining to, or concerned with, the stars or measured or determined by the stars.

Sidereal Day

The time needed for a complete rotation of the earth, measured as the interval between two successive meridian transits of the vernal equinox. The sidereal day is 23 hours, 56 minutes and 4.09 seconds in units of mean solar time.

Sidereal Hour

1/24 part of a sidereal day.

Sidereal Year

The time needed for one complete revolution of the earth about the sun – 365 days, 6 hours, 9 minutes and 9.54 seconds in units of mean solar time.

Siderite

A meteorite consisting principally of metallic iron.

Siderostat

An astronomical instrument that consists basically of a plane mirror rotated by a clockwork mechanism. The mirror reflects light from a celestial body in a relatively fixed direction to a fixed telescope for a long time. Also called a heliostat.

Siemens

The name of a famous family of Anglo-German engineers and inventors.

Ernst Werner von Siemens (1816–92), the founder of the dynasty, was born at Lenthe in Hanover, entered the Prussian army as a volunteer and studied at the Military Academy in Berlin. His interests included artillery, telegraph lines, electric traction, electroplating and dynamos. He was the first to use gutta-percha as an insulating material. He also founded the Physikalisch-Technische Reichsanstalt at Charlottenburg in 1886.

Sir William Siemens (originally Karl Wilhelm) (1823–83) brother of Ernst Werner (*q.v.*), also born at Lenthe and educated at the polytechnic school in Magdeburg and the University at Göttingen. He visited England at the age of 19 to try to sell the electroplating process he and his brother had developed. In 1844 he again went to England to market another 'chronometric' invention – a type of governor for steam engines. This time he decided to stay in England and became naturalized in 1859. His interests lay with the applications of heat and electricity and he constructed the Portrush electric tramway and many overland and submarine telegraphs.

Alexander Siemens (1847–1928) was born at Hanover and, when trained, went to

England and entered a firm at Woolwich in 1867. He assisted in the laying of the Indo-European telegraph line and the cable in the Black Sea; he was also responsible for the arc lighting in the Albert Hall and the British Museum.

Other members of the Siemens family were well known in Germany as engineers and inventors.

Slide Rule

An instrument which essentially consists of two logarithmically scaled rules mounted to slide along each other enabling division, multiplication and sometimes more complex computations to be carried out with ease. Most rules of this nature have a travelling cursor with a transparent panel and hair-line to assist with accurate readings of the results. Most rules are about 10 in long and by turning over the slide allow for further complex calculations. Varieties of slide rules include cylindrical versions with a central sliding sleeve, the Saxonic arithmometer that will give products to 12 figures, Thacker's cylindrical slide rule in fixed scales and Professor Fuller's 12 in rule with 3 in diameter and scales spirally arranged to a straight rule length of 83 ft.

The basic principle of slide rules was introduced by Professor Gunter (*q.v.*) of Gresham College, London, in 1620; Wingate added improvements in 1626.

Smeaton, John (1724–92)

English civil engineer who was born at Austhorpe Lodge near Leeds. At an early age he showed a talent for tool making and invention, producing a turning lathe when he was only 15. He was appreciated to a scientific instrument maker and then in 1750 set up in business on his own account making mathematical instruments which could be used in astronomy and navigation. His other interests included the making of water and wind mills, bridges and canal construction. He is best remembered for the rebuilding of the Eddystone lighthouse that had been burnt down in 1755.

Smith, Robert (1689–1768)

English mathematician who was born at Lea near Gainsborough and was educated at Leicester Grammar School and Trinity College, Cambridge. From 1716 to 1760 he was Plumian professor of astronomy. His writings included *A Complete System of Opticks, Harmonics* and *Harmonia Mensurarum*.

Smyth, Charles Piazzi (1819–1900)

British astronomer who was born at Naples, the name Piazzi was after his godfather, the Italian astronomer (*q.v.*). His father, Admiral Smyth, after leaving the service settled in Bedford where he built and equipped an observatory where Charles Piazzi received his first introductions to astronomy. When still only 16 he went to the Cape of Good Hope and assisted Sir Thomas Maclear with observations of Halley's (*q.v.*) comet and the great comet of 1843. In 1845 he was appointed Astronomer Royal for Scotland and professor of astronomy at the University of Edinburgh. In 1856 the Admiralty made him a grant of £500 and also lent a 140 ton yacht, the *Titania*, and a 7½ in equatorial telescope to go to Tenerife to investigate the possibilities of setting up experimental observatories on the top of high mountains, and to find out what the astronomical advantages could be. Charles Piazzi's findings confirmed Newton's (*q.v.*) thoughts that a 'most serene and quiet air ... may perhaps be found on the tops of the highest mountains above the grosser clouds.' Charles Piazzi's work included the use of the rain band in meteorology, spectroscopy and the construction of a map of the solar spectrum.

Solar Cell
A device that converts energy from the sun into electric energy.

Solar Eclipse *see* Eclipse

Solar Flare
A short powerful eruption of solar gases from a small area of the sun's surface. It is associated with a sunspot group.

Solar System
The sun, the nine planets and all other celestial bodies that orbit the sun. *See* Ptolemy *and* Copernicus.

Solenoid
A cylindrical device of insulated wire in which an axial magnetic field is established by a flow of electric current.

Solstice
Either of the two times in the year when the sun has no apparent northward or southward motion, at the most northern or most southern point of the ecliptic. The summer solstice is 21 June and the winter 21 December, coinciding with the longest and shortest day respectively.

South, Sir James (1785–1867)
British astronomer who was one of the

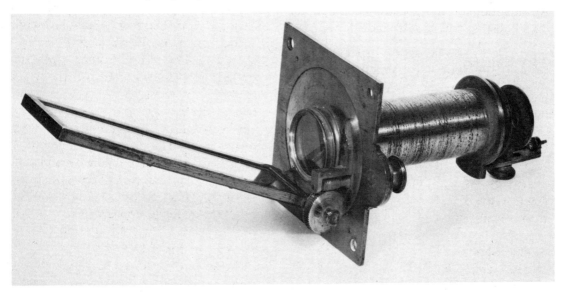

Brass solar microscope, the body tube mounted on an adjustable circular disc on the square mounting plate

founders of the Astronomical Society, of which he was elected President in 1829. He worked with Herschel (*q.v.*) in London, and with Laplace in Paris.

Specific Gravity

The ratio of the weight of any volume of a substance to an equal volume of a standard substance. The universally accepted standard is water at a temperature of 4 degrees Centigrade. The standard is sometimes varied in the comparison of certain gases.

Spectroheliograph

An instrument specifically for photographing the sun in the light which a particular element, usually calcium or hydrogen, emits. The resulting photograph is termed a spectroheliogram.

Spectrohelioscope

An adaption of the spectroheliograph for visual observation of solar radiation.

Spectrometer

A type of spectroscope which has scales for measuring the positions of the spectral lines.

Spectrophotometer

An instrument to determine the distribution of energy in a spectrum. The intensity signifies the abundance of the specific element causing the line.

Spectroscope

A collective term for a number of instruments for resolving and observing spectra. Early types included a glass prism; modern devices employ a diffraction grating.

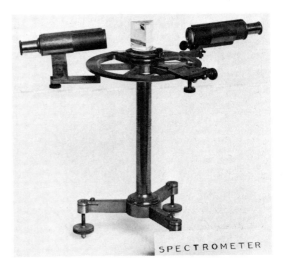

Mid-19th century spectrometer

Spectrum

The distribution of colours which results when white light is dispersed either through a prism or a diffraction grating. The study of spectra was probably begun by Sir Isaac Newton (*q.v.*). It is recorded that he bought a glass prism at Stourbridge Fair, near Cambridge. He then performed a series of experiments, which he set out in his *Opticks*. The wording of the first of these experiments was:

In a very dark Chamber at a round hole about one third part of an Inch broad made in the Shutter of a Window I placed a Glass Prism, whereby the beam of the Sun's Light which came in at that hole might be refracted upwards towards the opposite Wall of the chamber, and there form a coloured Image of the Sun.

The solar spectrum as noted by Newton would have been a continuous band of colour from red to violet. Fraunhofer (*q.v.*) found that the spectrum was crossed by numerous dark lines, which pointed to the fact that certain colours were missing from the complete spectrum. The spectrum of an element is individualized by a number of bright lines that are separated by small intervals; each element has its own specific and peculiar line spectrum. The spectra of gases such as helium, hydrogen and oxygen may be observed by passing an electric discharge through a Geissler (*q.v.*) tube that contains the chosen gas at a very low pressure. The end of the 19th century saw a great deal of research into the spectra and considerable progress was made in the science.

Spectrum Colours

Those visible in the spectrum of white light, in this order, are red, orange, yellow, green, blue indigo and violet.

Speculum

An alloy which, when polished, has a highly reflective surface; it was selected for the mirrors in the early reflector telescopes but has been replaced by silvered glass. The alloy was made principally from copper and tin, sometimes with other trace metals added to increase the 'whiteness'; some makers even included a small amount of arsenic. A typical speculum could consist basically of 126 parts copper with 58 parts tin.

Speed, John (1552–1629)

English cartographer and historian who was probably born at Farrington in Cheshire. His father was a London tailor and John for a time followed his father's trade, being admitted as a member of the Merchant Taylor's Company in 1580. His heart was not in the trade and he was fortunate in finding a patron, Sir Fulke Greville, whom he called 'procurer of my present estate' and said that from the generosity of Sir Fulke he also was rewarded with a 'waiter's room in the custom house', although what that might entail is not clear today. He put the leisure to pursue his antiquarian tastes to good use. In 1611 he brought out his *Theatre of the Empire of Great Britaine*; this consisted of 54 maps of different parts of the country which had appeared separately and with which he had been assisted by Christopher Saxton, John Norden and William White. In the same year Speed also brought out his *History of Great Britaine under the Conquests of the Romans ... to ... King James*. Generously, Speed acknowledged the help he had recieved from fellow antiquarians such as William Smith, John Barkham, Sir Henry Spelman and Sir Robert Cotton, who is thought to have revised proofs for him and made available manuscripts for study.

Sperry, Elmer Ambrose (1860–1930)

Extremely prolific American inventor born at Cortland, New York. He is probably best known for his application and adaption of the gyroscope whose properties he worked on from 1896. Although it is claimed that the gyrocompass should be credited to the German Anschutz-Kaempfe in 1908, the Sperry gyrocompass was fitted to the USS *Delaware* in 1910. From Sperry's inspired mind came forth at least the following: gyroscopic stabilizers for controlling the roll of ships in heavy weather, electric mining apparatus, electric locomotives, a high intensity arc searchlight, gyro-controlled torpedoes, methods for recovering tin from scrap food cans and a way of making caustic soda from salt.

Sphere

In strict geometrical terms, this is a three-dimensional closed surface with every point on the surface equidistant from the centre point.

In astronomy, the celestial sphere is the infinitely distant surface, part of which forms the dome of the sky on which we see the heavenly bodies projected. The eye of the observer can be taken as the centre and the variation of the observer's position does not affect the position of the surface because of its infinite distance.

Spherical Aberration

An optical defect that can occur with refracting and reflecting surfaces in which light rays from one axial point, incident on the surface at different distances from the optical axis, do not come to a common focus. This can be corrected either by polishing the optical surface to a non-spherical shape or by selecting optical components whose individual aberrations neutralize each other.

Spring Balance

An instrument for weighing which employs a helical spring. The object to be weighed is placed in a pan attached to the bottom of the spring and the extension moves a pointer which records the weight on a calibrated scale. It is termed a spring scale in America.

Spy Glass

A small telescope, or when plural binoculars.

19th century four-draw gilt-metal spy glass, the cushion-shaped base embossed and inset with turquoise

Stadia

A method of surveying distances with a telescopic instrument which has two parallel lines used to intercept intervals on a rod marked with calibrations. The intervals are proportional to the intervening distance. The parallel lines or cross-hairs are termed 'stadia hairs'.

Star

In astronomy, a self-luminous, self-containing roughly spherical body, a mass of gas in which the energy generated by nuclear reactions within is balanced by the outflow of energy to the surface and the inward-directed gravitational forces are balanced by the outward-directed radiation and gas pressures. A planet does not have sufficient internal heat to encourage nuclear reactions and thus its surface receives light from nearby stars.

Steelyard

A type of balance in which the body to be weighed is suspended from the shorter arm of a lever, which turns on a fulcrum, and a counterpoise slides along the longer arm, which is graduated to indicate the weight.

Stereophotogrammetry

The method of map making and surveying using stereoscopic pairs of photographs.

Stereoscope

An optical instrument which produces an illusion of depth or a third dimension in two-dimensional pictures.

Stop Watch

Small pocket watch which has a control to allow the seconds to be stopped or started as required. The earliest models date from the late 17th century but in these the action of stopping the seconds stopped the whole movement of the watch.

Strabo (c.63 BC–AD 25)

Greek geographer and historian who was born at Amasia in Pontus. He studied with Aristodemus, Tyrannio and Xenarchus and

From left: Brewster-type stereoscope; walnut stereoscope by Negretti and Zambra; Smith, Beck & Beck achromatic stereoscope; all English, late 19th century

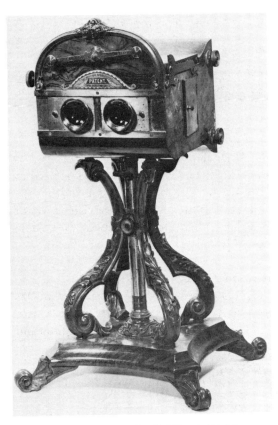

Natural stereoscope by J. Wood, Birkby, Huddersfield, c.1862

read Aristotle (*q.v.*). He travelled extensively and visited at least Italy, Egypt, Ethiopia and Sardinia. When in Egypt he spent some time in Alexandria where he could have had access to the works of such as Eratosthenes (*q.v.*) and Hipparchus (*q.v.*). It is not known exactly where his *Geographica* was written but what remains of the work marks it as the most important ancient record of the science of geography.

String Gnomon Dial
The term can be applied to any dial in which the gnomon is a stretched cord or thread.

Stroboscope
An instrument used to determine speeds of rotation or vibration.

Struve
A family of German astronomers.
Friedrich Georg Wilhelm (1793–1864) was born at Altona. He entered the University of Dorpat where he at first studied philology and then turned to astronomy. He became extraordinary professor of astronomy and finally director of the Observatory; he stayed in Dorpat studying double stars and geodesy until 1839 when he left to supervise the construction of a new central Observatory at Pulkowa, near what was then St Petersburg.
Otto Wilhelm (1819–1905), Friedrich's (*q.v.*) elder son, studied in the academy at St Petersburg and then became director of Pulkowa when his father resigned; he was also advisor on astronomy to the army and navy. His particular interests were double stars, of which he discovered some 500 and Saturn and its rings.
Heinrich Wilhelm (1822–c.1900), another of Friedrich's (*q.v.*) sons studied chemistry and fill'd a number of important administrative posts.
Karl Herman (1854–1920), one of Otto's (*q.v.*) sons, also studied at Dorpat and followed his father to Pulkowa. He then went to Königsberg Observatory and finally the Berlin Observatory.
Gustav Wilhelm Ludwig (1858–c.1920), also Otto's (*q.v.*) son, became observer at the Dorpat Observatory.

Sun

The centre of the solar system and the source of heat and light for the planets in this system. The sun has a mean distance from the earth of about 93 million miles (149,000,000 km), a diameter of around 864,000 miles and a mass about 330,000 times that of the earth.

As this great fiery wonder has been the support for life so have the early and primitive peoples regarded it as a leading deity, under a variety of names. The principal sun god of the Assyrians was Shamash, Merodach of the Chaldeans, Ormuzd of the Persians, Ra of the Egyptians, Tezcatlipoca of the Mexicans, Helios of the Greeks and Sol of the Romans. Helios was seen as driving his chariot daily across the heavens, coming up out of the sea at dawn and sinking back into it at evening. The names of his coursers were Brontë, representing thunder, Eoos, for daybreak, Ethiops, for flashing, Ethon the fiery, Erythreos, the red-producer and Philogea, the earth loving. The Scandinavians had a sun god, Sunna, who was in continual dread of being devoured by a wolf called Fenris; this symbolized an eclipse. Sunna was carried through the skies by the great horses Arvakur, Aslo and Alsvidur.

Sundials

Instruments which indicate the time of day during sunlight hours. A stationary angled arm, termed a gnomon, casts a shadow on to a plate marked with the hours. *See* Dials.

Sunspot

Any of the relatively dark spots that can appear in groups on the surface of the sun.

The depth of tone indicates a loss of heat; the surface material loses energy whilst at the centre of a high magnetic field. The appearance of sunspots varies within a period of about 11 years.

Superior Planets

Jupiter, Mars, Neptune, Pluto, Saturn and Uranus.

Supernova

A rare celestial phenomenon, this is the explosion of a gigantic star, which can be considerably larger than the sun. The immense explosion is caused by the available stock of hydrogen being exhausted very quickly. For a brief period, possibly a few days, it emits a fantastically bright light as the core temperature rises to hundreds of millions of degrees. Then there is a collapse and a vast release of energy carries most of the star's body out into space.

Surveying

The science of attaining a mathematically accurate representation of the relative points and positions of features on the earth's surface.

Surveyor's Chain

A measuring instrument of wire or metal strip links that is 22 yards long; each link measures 7.92 in. It is suitable for land measurement, 10 square chains equalling one acre. It is also called Gunter's (*q.v.*) chain.

Swan, Sir Joseph Wilson (1828–1914)

English physicist and electrician who was born in Sunderland. He is best recalled for his incandescent filament electric lamp, introduced in 1879, which was the first successful one of its kind. His other inventions and developments included a miner's electric safety lamp, the carbon process of printing in photography and the 'rapid' plate process, also in photography.

Synodic Period

In astronomy, this the interval of time between two successive conjunctions of the same kind with planets and the sun; with satellites it is the period relative to the radius vector from the sun.

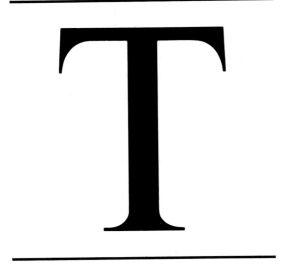

Table Clock

A clock that has a horizontal dial and is designed to be viewed from above.

Tables

In astronomy, the documentation of the movements of the celestial bodies during set periods of time. They were studied at length by many of the early astronomers and much work was done to constantly revise and correct those by earlier hands. Editions of these include Alfonsine, Rudolphine and Toledan. In navigation, tables are concerned with tidal movements, giving high and low waters, spring and neap tides at various set places around the world to give maximum assistance to the navigator. In mathematics, the term signifies tables of logarithms, trigonometry, fractions, etc.

Tacheometer

The name given to instruments, of which there are a number of variations, for the rapid location of points when surveying, both horizontally and vertically. They can be particularly useful when the ground is hilly or obstructed by trees and bushes. The tacheometer would be used with a pole similar to a level staff; this would be marked with heights from the foot and graduated according to the type of tacheometer being used. The horizontal distance is calculated either from the vertical angle included between two well-defined points on the staff and the known distance between them, or by readings of the staff indicated by two fixed wires in the diaphragm of the telescope. The difference of height is computed from the angle of depression or elevation of a fixed point on the staff and the horizontal distance already obtained.

Tachometer

A device for measuring the speed of revolution of a shaft. The unit used is revolutions per minute (rpm).

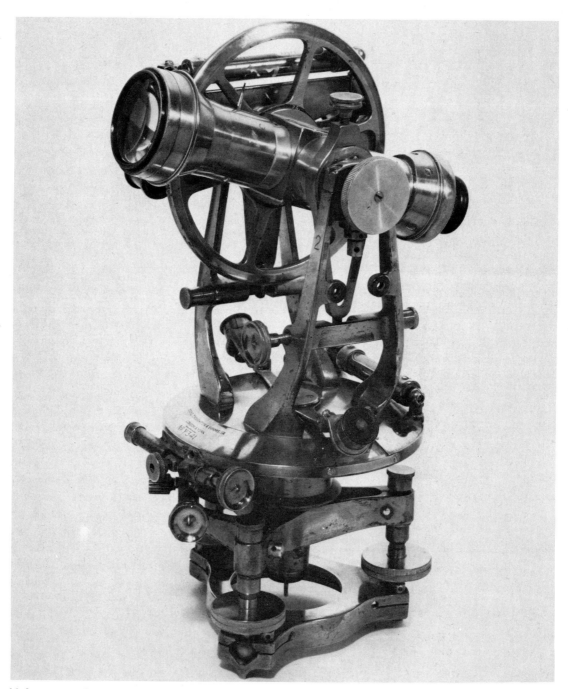

19th century brass tacheometer by Cooke, Troughton & Simms, London, with vertical and horizontal circles, the telescope having a Ramsden eyepiece

Tait, Peter Guthrie (1831–1901)

Scottish physicist and mathematician who was born at Dalkeith and educated at the Edinburgh Academy. His principal researches were into quaternions, thermodynamics and thermo-electricity. He collaborated with Balfour Stewart in writing *The Unseen Universe* and *Paradoxical Philosophy* and with W. J. Steele in *The Dynamics of a Particle*.

Talbot, William Henry Fox (1800–77)

English photography pioneer who was born at Lacock Abbey in Wiltshire and educated at Harrow and Trinity College, Cambridge. He worked principally in mathematics and optics, and was interested in the chemical changes of colour. He developed the calotype process of photography. He noted his photographic findings in his *Pencil of Nature* published in 1844.

Telautograph

An instrument for transmitting sketches or written messages by telegraphy. The sketch or message is drawn or written with a pencil on a roll of paper. The movement of the pencil is resolved into its component rotary motions and these control currents in two separate circuits. The signals are received and translated back by a recorder which sets in motion another pencil to express the sketch or message.

Telegraph

Any communication system that transmits and receives messages over a distance by using either radio signals or electrical impulse signals sent along a transmission line to a receiving or transmitting station.

Modern telegraphy had its origin in Oersted's discovery in 1819 of the magnetic field produced by an electric current.

Telephone

An instrument that directly modulates the sound waves of the voice or other acoustic source enabling them to be transmitted to far distant locations where they are reconverted into audible and understandable signals. The system, now worldwide, relies on direct wire connection or a combination of wire and radio signal transmission.

The term 'telephony' was first used by Philipp Reis of Friedrichsdorf in a lecture to the Frankfurt Physical Society in 1861. It is difficult to pinpoint who was responsible for the basic idea; it seems that the telephone's history was a progressive development by a number of highly alert scientific minds. In 1831 Wheatstone (*q.v.*) carried out experiments with the sounding boards of two musical instruments connected by a rod of pine wood. This showed that notes played on one could be transmitted across the rod to the second instrument which would truthfully reproduce the sounds. In July 1837 Dr Page of Salem, Massachusetts, noted that there was a sound made at the moment when the electric circuit to an electromagnet was broken. In the 26 August 1854 edition of *L'Illustration* Charles Bourseul wrote about flexible plates that would vibrate in sympathy with varying pressures of air and therefore, open and close an electric circuit. Underlying a great deal of inventive thought was the pioneer work of Michael Faraday (*q.v.*) in the field of electromagnetism. If the accolade can be awarded to one person this must be the Scottish-American physicist Alexander Graham Bell (*q.v.*) who first satisfactorily solved the problem of being able to convert sound energy into electrical energy at one station and then reconvert the electrical energy back into sound energy at the receiving station.

The possibility for worldwide communication through this new medium brought a spate of commercial activity and much work for the lawyers handling the requests for protective patents for the various inventors. In the United Kingdom alone some of the facts illuminate the fever that must have run high in the latter part of the 19th century. In 1876 a patent was granted for Graham Bell's system and in 1877 a patent was granted for Edison's (*q.v.*) system. In 1878 Professor Hughes invented the microphone but did not apply for a patent. That same year, the Telephone Company was formed to take over Bell's patent. In 1879 the Edison Telephone Company was formed; this merged with the Telelphone Company to become the United Telephone Company Limited in 1880. Arguments arose as to whether or not the telephone operation should be handled by the Post Office, which issued some limited distance licences in 1881. Some 70 companies were now in the field bidding for what they felt must be the sound equivalent of a goldrush. In 1883 the Post Office tried to enter the competition but the Treasury was opposed. Confusion grew over the rights to lay lines and operate and in 1896 trunk wires were transferred to the Post Office, which was allowed to develop its telephone exchange business by the 1899 Telegraph Act.

Telescope

An optical instrument which, by a combination of lenses, can make distant objects appear closer. The credit for the discovery of the telescope principle must lie with Lippershey and Galileo (*q.v.*), although there are grounds to believe that it may have been an earlier invention. Democritus (*c.*460BC–*c.*370BC), the Greek philosopher who elaborated on the atomic theory expressed by his contemporary Leucippus, pronounced that the Milky Way was composed of hosts of stars. Some people take this as evidence that he must have had some means of examining the heavens. William Molyneux in his *Dioptrica Nova* of 1692 thought that Roger Bacon, who died in 1294,

...did perfectly well understand all kinds of optic glasses, and knew likewise the method of combining them so as to compose some such instrument as our telescope.

Further, he points to a passage in Bacon's *Epistola ad Parisiensem* 'Of the Secrets of Art and Nature' (Chapter 5):

Glasses or diaphonous bodies may be so formed that the most remote objects may appear just at hand, and the contrary, so that we may read the smallest letters at an incredible distance, and may number things, though never so small, and may make the stars also appear as near as we please.

Certainly such passages indicate that Roger Bacon may have arrived at the theory but there is no mention of him actually trying the ideas out in practice. In 1558 Giambattista della Porta made this statement in his *Magia Naturalis*:

18th century ivory monocular, outer barrel carved and pierced

If you do but know how to join the two [concave and convex lenses] rightly together, you will see both remote and near objects larger than they otherwise appear, and withal very distinct.

In 1579 in his *Stratioticus* Thomas Digges claimed that his father, Leonard Digges,

... among other curious practices had a method of discovering by perspective glasses set at due angles all objects pretty far distant that the sun shone upon, which lay in the country round about.

Leonard Digges' own publication, *Pantometria,* published by his son in 1571, contained the passage:

Marvellous are the conclusions that may be performed by glasses concave and convex, of circular and parabolic forms, using for mutliplication of beams sometimes the aid of glasses transparent, which, by fraction, should unite or dissipate the images or figures presented by the reflection of other.

Evidence such as the foregoing sheds a fascinating light on the workings of an underlying creative source that the minds of thinkers pick up, sometimes unconsciously. The process occurs repeatedly with receptive inventive individuals happening on the same idea though they may be centuries apart and geographically distant from each other.

See Cassegrain, Gregorian, Reflector, Refractor, Collimator, Galileo, Newton, Hevelius, Huygens, Herschel and Rosse.

Terrestrial Globe
One on which the map of the world is projected, represented and drawn.

Tesla, Nikola (1857–1943)
Austro-Hungarian-Yugoslav inventor. His birthplace is a little uncertain, but he emigrated to the United States in 1884. Tesla was associated with Edison (*q.v.*) and the Westinghouse Company and later had his own laboratory. His brilliant mind encompassed the development of alternating current with generators and motors and opened the way for long distance transmission of alternating current for industrial and domestic use. His inventions and developments included the Tesla coil, Tesla tube, oscillators and electric lamps and many other electric appliances and devices.

30 in terrestrial globe by W. & A. K. Johnston Ltd, showing the submarine cable lines

Thales (640–546 BC)

Greek philosopher and scholar who is recognized as the founder of Grecian astronomy, geometry and philosophy. He was born at Miletus in Asia Minor, and respected by his contemporaries as chief of the seven wise men of Greece. One of his noteworthy predictions was of the eclipse which took place on 28 May 585 BC. This act was recorded by Herodotus who accounts that it took place during a great battle between the Lydians and the Medes and that the sudden happening stopped the hostilities and brought a lasting peace between the contestants. He founded the Ionian school of natural philosophy, and claimed that water was the all-pervading principle of the universe and that from it all matter was formed. He developed thoughts on geometry culled from earlier workers from Egypt and ideas on astronomy from the Chaldeans.

19th century pocket terrestrial globe published by Newton Son & Berry, 66 Chancery Lane, London. The inner concave surface is printed with a celestial globe showing constellations, heavenly bodies, mythical figures and instruments.

Theodolite

One of the most useful instruments in surveying. It measures both vertical and horizontal angles. It consists of a telescope mounted so as to move on two graduated circles, one vertical and the other horizontal. The axes of the telescope pass through the centre of these two circles. The earliest instruments based on this principle were Islamic, and it is likely that the first example in Europe was Martin Waldseemüller's polimetrum. Before the beginning of the 18th century, theodolites used open sights. The first true theodolite as known today was

Brass transit theodolite by Troughton & Simms, late 19th century

Sisson-type theodolite, English, late 19th century

made about 1720 by Sisson. Benjamin Cole, working around 1765 from a business address of 'Cole, maker at ye ORRERY in Fleet Street, LONDON' equipped some theodolites with two spirit levels which are essential for establishing the instrument horizontally. Jesse Ramsden (*q.v.*) was making geodetic theodolites, instruments which could take into account the curvature of the earth, at the end of the 18th century.

Theodosius (active in the 1st century BC)

Greek astronomer, sometimes incorrectly named Theodosius of Tripolis. He was actually a native of Nithynia and is renowned for his works on geometry, the best known being *Sphaerica* which deals with pure spherical geometry. By repute, he developed the sundial.

Thermograph

A thermometer which keeps a continuous record of the temperatures prevailing in its habitat.

Thermometer

An instrument for measuring temperature. It usually consists of a glass tube with a bulb containing mercury or, in special cases, a liquid that will react to temperature changes by expansion or contraction. The sides of the glass tube or the mount on which the tube stands may be scaled. The clinical thermometer is specifically for measuring the temperature of the body. A maximum and minimum thermometer records the highest and lowest temperatures in a given time span.

Thermostat

A device which automatically reacts to temperature changes and activates switches controlling boilers or refrigerators, for example.

Thomson, Elihu (1853–1937)

English electrician and inventor who was born in Manchester and emigrated to America with his parents whilst still a child. He was educated at the Central High School in Philadelphia and became professor of mechanics and chemistry there. With E.J. Houston he founded the Thomson-Houston Electric Company, which became part of General Electric Company. Thomson was a prolific inventor, finally holding some 600 patents for devices of all kinds associated

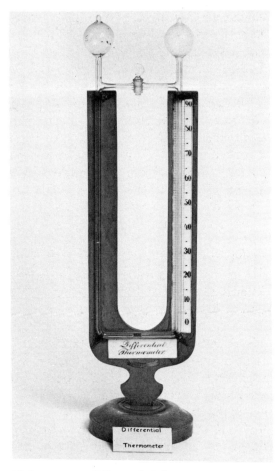

19th century differential thermometer

with electricity; he also made improvements to electric arcs and welding apparatus, for example. He was the first person to make stereoscopic X-ray photographs.

Thomson, James (1822–92)

British physicist born in Belfast, the elder brother of Lord Kelvin (*q.v.*). His principal researches were in the fields of thermodynamics, liquefaction of gases, water wheels and water turbines.

Thomson, Sir Joseph John (1856–1940)

English physicist, born near Manchester and educated at Owens College and Trinity College, Cambridge. He is credited with the discovery of the electron and his other important work included research on cathode rays, magnetic fields, conduction of electricity through gases and the electrical theory of inertia of matter. His books included *Elements of the Mathematical Theory of Electricity and Magnetism* and *The Electron in Chemistry*.

Tides

Regular and predictable movements of the oceans produced by the action of the gravitational forces of the sun and moon. The lunar effect is the more powerful.

Tisserand, François Felix (1845–96)

French astronomer, born at Nuits-Saint-George on the Côte-d'Or and educated at the Ecole Normale Supérieure. In 1866 he joined the Paris Observatory and was part of the team that went to Malacca to observe the solar eclipse of 18 August 1868. In 1873 he was appointed director of the Observatory at Toulouse, and in 1892 director of the Paris Observatory. His most important work was *Traité de Mécanique Céleste* which ranks beside *Mécanique Céleste* by Laplace. His principal researches were in mathematical astronomy.

Tompion, Thomas (1639–1713)

English clockmaker, born at Northill, Bedfordshire. He became one of the world's greatest craftsmen at his trade. Sadly, probably fewer than 20 of his more celebrated constructions have survived but his perfectionism had a great influence on those working around him. When he came to London he worked first in Blackfriars and later was to be appointed as clock and watchmaker to the Court of Charles II. Praise came from the Continent and he fully deserved the title of 'father of English watchmaking'.

Torricelli, Evangelista (1608–47)

Italian physicist and mathematician who was born at Faenza, and having been left fatherless at an early age he was educated under the care of an uncle, a Camaldolese monk who in turn sent him to Rome to study under the Benedictine, Benedetto Castelli, professor of mathematics at the Collegio di Sapienza. Later he acted as Galileo's (*q.v.*) secretary. The influence of the great man fired Torricelli himself to research, and the results included the principle of the barometer, the movement of bodies and their centre of gravity, parabolas of projectiles and improvements to microscopes and telescopes.

Transit

In astronomy, the passage of a celestial body across the observer's meridian.

Transit Circle

A telescope that is mounted firmly so that it can only swing in a vertical, north–south, plane. The instrument can therefore be used to time and observe the transit of a celestial body across a meridian.

Traverse Board

A board with the four points of the compass marked on it. For each point eight holes are bored, one for each half hour of the watch when at sea. It was formerly used to record the courses made by the ship in each half hour.

Trepidation

A supposed oscillation of the equinoxes. The word is used technically for an imagined slow oscillation of the ecliptic. It was introduced by the Arabian astronomers to explain a possible variation in the precession of the equinoxes. The idea appeared in astronomical tables until the time of Copernicus (*q.v.*). It was based on an error in Ptolemy's (*q.v.*) determination of precession.

Trigonometry

The branch of mathematics that is primarily concerned with the measurement of plane and spherical angles. Its scope includes all types of geometric and algebraic problems. The science is deeply rooted; Hipparchus (*q.v.*) invented theorems and Ptolemy (*q.v.*) dealt with aspects of the subject in his *Almagest*, the Indian astronomer Aryabhata (476–550) discussed branches of the thought in a long poem. The Arab astronomer Albategnius, who died in about 929, used aspects of the science and Copernicus (*q.v.*) gave the first simple demonstration of the fundamental formula of spherical trigonometry.

Tyndall, John (1820–93)

Irish natural philosopher who was born in Co. Carlow, the son of a lowly landowner. His career was various: helping with the ordinance survey in Ireland and studied chemistry under Bunsen in Marburg. His researches included radiant heat and the acoustic properties of the atmosphere. His most important publication was *Contributions to Molecular Physics in the Domain of Radiant Heat*. In 1866 he became scientific adviser to Trinity House and carried out important experiments on sound for them.

Ultramicroscope

An instrument with a high-intensity illumination which is used to examine very minute objects. The illumination generally comes in from the side so the particles stand out against a dark background.

Ultraviolet Light

This is the portion of the electromagnetic spectrum which has wavelengths longer than X-rays but shorter than light; it is just beyond violet in the visible spectrum. Although it is invisible to the naked eye, it can be seen by the fluorescence it causes when allowed to play over certain substances, which in their turn are also invisible to the human eye before the ultraviolet light falls on them. The rays reach the earth in considerable quantities from the sun, though much of the outflow is prevented from reaching the earth by the ozone layer in the upper atmosphere.

An ultraviolet lamp is one that is filled with mercury vapour and produces ultraviolet light artificially. The lamp can be very useful for examining various categories of antiques and paintings; comparatively recent retouching and the use of adhesives and lacquers with restoration will fluoresce.

Ultraviolet Microscope

This has quartz lenses and the object is recorded photographically; illumination is by ultraviolet radiation.

Ulugh Beg, Mirza Mahommed Ben Shah Rok (1394–1449)

Persian astronomer, the son of Shah Rok and grandson of Tamerlane, he became Prince of Samarkand in 1447 but was murdered by his eldest son two years later. He is famous for building an observatory at Samarkand, from which tables of the sun, moon and planets were issued. His introduction to these tables is of particular interest as it shows clearly the astronomical and trigonometrical methods that were in use at the time. Ulugh Beg also found serious errors in the earlier Arabian star catalogues, which in most cases were simply copied from Ptolemy (q.v.), with the effect of precession added to the longitudes; he redetermined the positions of 992 fixed stars and also added 27 stars from Al Sûfi's catalogue, although these were too far south to be observed from Samarkand. The introduction to his tables was translated into French as *Prolégomènes des tables astronomiques d'Ouloug Beg* in 1853.

The Arabians deserve much of the credit for keeping the study of the heavens alive during the Dark Ages. They built some excellent observatories in Spain and in the area of Baghdad. The Mongol emperors of India were also enthusiastic and erected some large astronomical instruments in the most important cities of their empire.

Universal Ring Dial *See* Ring Dial

Universal Time

This is another name for Greenwich Mean Time (*see* Greenwich Observatory) and is a system for reckoning time uniformly throughout the world for international purposes. The system was agreed upon at an International Conference in Washington in 1883. By it the day is taken as having 24 hours; the circumference of the earth is divided into 24 parts of 15 degrees each and a local time is fixed in each case. Odd minutes and seconds are ignored so that the various local times will differ from Universal Time by only even hours. All of this facilitates the alteration of clocks on trans-world transport and international signalling.

Universe

In astronomy, all space, existing energy and matter, the cosmos, the macrocosm. Over the centuries the term has had to move outward as man's reflected creativity discovered more powerful instruments and new ideas pushed limiting boundaries out towards infinity. Ptolemy (*q.v.*) put forward a theory which today seems cramped, almost bound by its conception. Eudoxus (*q.v.*) suggested a universe composed of twenty-seven concentric spheres with the earth as the common centre. Archimedes' (*q.v.*) free thinking led him to a counting system that could handle the huge numbers needed to calculate the immensity of the universe. Copernicus (*q.v.*) developed a theory of the universe with the sun as the centre of the solar system. Radio astronomy has moved into distant reaches of space that could have only been guessed at by earlier astronomers. Some of those surveying the results of observation must be moving towards a conviction that out there past the comparatively nearby shining heavenly bodies lies a path that has no ending: the universe can be recognized as a synonym for the infinite.

Ursa Major (the Great Bear)

One of the most conspicuous constellations in the northern hemisphere and one that can be found quite easily, for it is never below the horizon in Britain. As with many other large celestial bodies, Ursa Major attracted its share of mythology and legend. The most popular legend was that Calisto, daughter of Lycaon, was violated by Jupiter. Juno changed her into a bear, and Jupiter placed her among the stars that she might be more under his protection. The constellation is supposedly referred to in *Job* 9, v. 9; Homer called it Arktos, the Bear, and Hamaxa, the Wagon and Eudoxus (*q.v.*) referred to it, as did Aratus. The Romans called it Ursa, the Bear, and Sep-

temtriones, the seven ploughing oxen. Ptolemy (*q.v.*) catalogued eight stars, Tycho Brahe (*q.v.*), seven and Hevelius, twelve. Other names include the Plough and the Dipper.

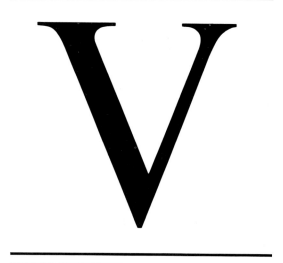

Variable Star

One which varies considerably in brightness, possibly from internal changes or by periodic eclipsing. One of the first of these to attract attention was Omicron Ceti in the constellation of Cetus, which was observed by Fabricius in 1596. Another was the Demon, Al gûl, so-called by the early Arabians, which was noted by Goodricke in 1783.

Vernier, Pierre (1580–1637)

French inventor, who was born at Ornans in Burgundy and spent some of his early years as commandant of the castle in his native town. He invented the device that bears his name which allows the reading of the fractions of the smaller parts of a measuring scale. It is used on all instruments which take linear or angular measurements such as barometers, cathetometers, theodolites, sextants and certain telescopes, etc. It is sometimes called a nonius, especially in Germany, after Pedro Nuñez (1492–1577), professor of mathematics at Coimbra University, but this is incorrect.

Vertical Dial

An instrument in which the hour scale is vertical.

Viscometer

A device for measuring the viscosity of a liquid.

Vogel, Herman Karl (1841–1907)

German astronomer who was born at Leipzig. After holding the post of assistant observer he became director of the Astrophysical Observatory at Potsdam in 1882. His most important work was his investigations on the spectrographic determination of the radial morion of the stars. He published the first spectroscopic catalogue of the stars in 1883.

Volt

The international system (SI) unit of electric potential and electromotive force; it is named after Count Alessandro Volta (*q.v.*).

Volta, Count Alessandro (1745–1827)

Italian physicist, born at Como who became famous for his discoveries in electricity. One time professor of natural philosophy at Pavia University, his discoveries and re-

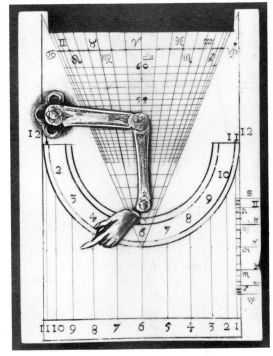

Ivory universal altitude dial, probably French late 16th/early 17th century, after the designs of Regiomontanus published in 1476. To find the time the finger of the indicator is adjusted to the latitude and date at which the dial is to be used. The bead on the plumb-bob is then adjusted to the date on the small zodiac scale and the dial is held up until the light passing through the upper sight falls upon the lower sight, the plumb-bob showing the time on the vertical hour lines.

search included the electrophorus, an electrical condenser, the hydrogen lamp, the development of electricity in metallic bodies and the 'Voltaic' pile. He received the Copley Medal from the Royal Society and

Napoleon had a special medal struck in his honour. He was a friend of Buffon, Lavoisier and Galvani.

Voltameter
Another name for a coulometer (*see* Coulometer).

Voltammeter
An instrument intended to measure current or potential.

Wag-on-the-Wall
An American slang term for the dial and works of a longcase clock that were hung on a wall without the case. This was sometimes done whilst the clock maker was still working on a particular instrument.

Water Clock *see* Clepsydra

Watt
The practical unit of electrical power. The number of watts is obtained by multiplying the numbers of volts and amperes which are operating in the specific case.

Waywiser
Also called an odometer, perambulator or surveyor's wheel, this is basically a large wheel linked to a dial, that can record the number of revolutions, and thus a reasonably accurate measurement, of the distance between places. Probably the earliest mention of such instruments is in the writings of Vitruvius, who described a gadget with cogs that could be connected to a carriage wheel and record the number of revolutions. Metal wheeled modern versions may be seen today being trundled along a road by a junior member of some council's surveying staff.

Wegener, Alfred (1880–1930)
German meteorologist who became departmental director of the German State Marine Meteorological Institute at Hamburg, and later was professor of meteorology at Graz. He carried out a series of scientific expeditions to Greenland, and on the last of these, when trying to cross the icecap in midwinter, he died. He held to a theory that continents are all adrift, and that the earth originally was a compact mass and that this had cracked and the different portions had gradually drifted apart. His written works included *Thermodynamik der Atmosphäre* and *Windun Wasserhofen in Europa*.

Westinghouse, George (1846–1914)
American inventor and industrialist who

schemes for the long-distance transmission of alternating current, in conjunction with Nikola Tesla (*q.v.*).

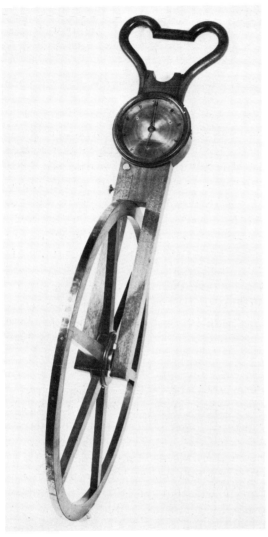

19th century waywiser by Thos Rubergall

Wet and Dry Bulb Hygrometer

An instrument for determining the relative humidity of the atmosphere. It consists in the main of a pair of thermometers mounted side by side; the bulb of one is covered by a small piece of muslin that should be kept moistened. The one with the muslin records a lower temperature than the other, owing to the heat loss by evaporation. It may also be called a psychrometer.

Wheatstone, Sir Charles (1802–75)

English physicist who was born at Gloucester, the son of a music seller. His education came from a succession of private schools who recorded that he showed little talent and that he was morbidly shy and withdrawn. In 1816 his father sent him to a musical instrument maker in the Strand, London, to learn the trade, but with his father's permission he set about a programme of reading that was to educate him far more than any of the schools. In 1834 he was made professor of experimental physics at King's College, London. He still remained so shy that he was unable to lecture to a large group and it was Faraday (*q.v.*) who described many of his investigations in his Friday evening discourses at the Royal Institution. The ground covered by this strange shy person was impressive: the prismatic analysis of electric light, the transmission of electric signals over wires, the magneto-electric dial telegraph, the type-printing telegraph and automatic transmitting and receiving instruments. His most

was born at Central Bridge, New York. During his inventive life he filed more than 400 patents for such ideas as a device for re-railing derailed steam cars, his celebrated air brake, pneumatic points control, switching and signalling systems and long-distance natural gas pipelines. He also worked on

famous achievement was the Wheatstone Bridge, an instrument for measuring an unknown electrical resistance.

Wheel Barometer

A kind of siphon barometer which shows the mercury level magnified and registered on a dial.

White Dwarf

A small faint star with a very high density, which is known to cause a specific shift of spectral lines, thought to mark the final stage in the evolution of the stars. As its material collapses gravitationally, the cooling process starts until it loses luminosity. It is then known as a black dwarf.

Wild, Heinrich von (1833–1902)

Swiss meteorologist who was born at Uster in the canton of Zürich; in 1858 he became director of the Observatory at Bern which he made into a central meteorological bureau. His services for such work were in demand and he repeated the effort in Russia, establishing Observatories at Pavlovsk and Irkutsk. His inventions included a saccharometer and a magnetic theodolite. His most useful publication was *Temperaturverhältnisse des russischen Reichs.*

Window Dial

One that is painted on a glass sheet for inclusion in a window; for observation purposes it has to be a vertical dial.

Wolf, Max Franz Joseph Cornelius (1863–1932)

German astronomer who was born at Heidelberg. His appointments included director of Königstühl Observatory and professor of astrophysics at Heidelberg University. He discovered a periodical comet in 1884, which was named after him and he did important work with the development of celestial photography, during which he discovered more than 200 asteroids.

Wollaston, William Hyde (1766–1828)

English chemist and natural philosopher who was born at East Dereham, Norfolk, the second of 17 children. His father, the Rev. Francis Wollaston, was a keen amateur astronomer. William was educated at Charterhouse and Caius College, Cambridge and then set out on a career in medicine, but, after failing to obtain a post at St George's Hospital, he abandoned the idea of being a doctor and turned to his own original research which included detailed experiments with optics, which led to his discovery of the dark lines in the solar spectrum; his invention of a reflecting goniometer and a form of the camera lucida; trials with the use of concavo-convex lenses for the oculist; the effects of an electric current on a magnetic needle; and the finding of a way of making platinum malleable. He was also the first to detect the metals palladium, rhodium, columbium and titanium.

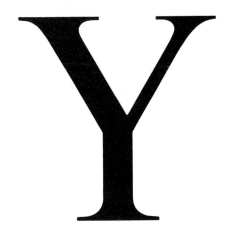

Xerography

A dry photographic or photocopying technique in which a negative image made by a resinous powder on an electrically charged plate is electrically transferred to and thermally fixed on paper.

X-ray Crystallography

The examination of the structure of crystals by passing a beam of X-rays through them and studying the diffraction patterns.

X-rays

Electromagnetic radiation with wavelengths between gamma radiation and ultraviolet radiation.

Yerkes, Thomas Tyson (1837–1905)

American financier, born in Philadelphia, whose fortunes were largely based on his development of the street railway system of his native town and the Chicago elevated railway, also his work with the Tube system in London. A noted philanthropist who, amongst other gifts, endowed the Chicago Observatory, including the 101 cm telescope which was installed at Lake Geneva about 80 miles to the north west of the city. This huge instrument, a refractor, with a focal length of about 62 ft, has given many accurate observations of star parallaxes. Yerkes was also a connoisseur with an outstanding collection of paintings, tapestries and rugs, which was sold in 1910 for $2,034,450.

Young, Charles Augustus (1834–1908)

American astronomer who was born at Hanover, New Jersey, and became professor of natural philosophy and mathematics at the Western Reserve College, Ohio. He is principally remembered as the discoverer of the reversing layer in the atmosphere of the sun and for being the first to observe the spectrum of the solar corona.

Young, Thomas (1773–1829)

English physicist and Egyptologist who was born at Milverton, Somerset. He had an amazing talent for languages which enabled him to make a valuable study of the writings of those from many lands. By the time he was 14, he was acquainted with Greek, Latin, French, Italian, Hebrew, Persian and Arabic. One use he made of this was to translate the demotic text from the Rosetta Stone. Young is best recalled for his work in physical optics and for being the author of a number of treatises which helped to establish the undulatory theory of light and the interference of light. He worked on from Newton's (*q.v.*) theories of light and Huygen's (*q.v.*) wave theory.

Zadkiel (1795–1874)

The pseudonym of Richard James Morrison, the English astrologer, whose early life was spent in the coastguard service. He founded the *Herald of Astrology*, later known as *Zadkiel's Almanac*, which achieved an enormous circulation. He was also the author of a number of highly controversial books on astronomy.

Zeeman, Pieter (1865–c.1940)

Dutch physicist who was born at Zonnemaire and later became director of the physical institute at Amsterdam. He was an expert on magneto-optics and was the discoverer of the Zeeman effect, which concerned the spectral lines in a magnetic field.

Zeiss, Carl (1816–1888)

German optician who was born at Weimar and started out by being apprenticed to a number of instrument makers in Weimar, Stuttgart and Vienna. In 1846 he opened his own business in Jena. Later he worked with Otto Schott, founder of the Jena Glass Works, on microscopes and on perfecting the homogeneous immersion lens. The Zeiss firm has prospered with its work on all kinds of optical instruments.

Zenith

The point in the sky which is directly overhead. The opposite of the nadir, which is the point on the celestial sphere directly below the observer.

Zodiac

A belt of the celestial sphere 18 degrees wide, extending for 9 degrees on each side of the ecliptic. As most of the planetary orbits are almost in the same plane as the earth's, the planets are always to be found in the Zodiac, except for Pluto which has periods of absence. The study of and interest in the Zodiac dates right back to the beginnings of astronomy and also very much astrology. The Zodiac was split into 12 divisions of 30 degrees, each with its own sign: Aries, Taurus, Gemini, Cancer, Leo, Virgo, Libra, Scorpio, Sagittarius, Capricorn, Aquarius and Pisces. The ancient peoples gave seasonal significance to the phases, although names amongst the Chinese, Hindus, Chaldeans, Egyptians and Greeks often differed. In the past there was often much confusion as astronomy and astrology became intertwined, particularly in this area of study.

Zodiacal Light
A faint haze of light with a conical shape which extends from the sun along the ecliptic and is visible just after sunset and just before sunrise. It may be caused by the reflection of sunlight from cosmic dust in the plane of the ecliptic. One of the best times to observe this phenomenon is in the evening about the time of the vernal equinox.

APPENDICES

CARE OF THE COLLECTION

The problem is perhaps a little more complicated than might first be thought. Scientific instruments can be constructed from a number of different substances; apart from alloys and pure metals, they may be covered with animal or fish skins, and the mounts and cases are likely to be of wood. There can be detailed parts made from glass, bone, ivory, and also applied painted and engraved decoration. The handling of them calls for care and understanding.

An important factor is that the environment for them should as far as possible be stabilized. An over-damp atmosphere should be avoided. The ideal would be to achieve a relative humidity of between 55 and 60 per cent, and coupled with this there should be no excessive heat overall throughout the building, as an excessively dry atmosphere affects leather, ivory, bone and wood. A careful eye must be kept to see that no specimen is unduly close to a source of heat such as a radiator or electric or other fire. An instrument called a hygrometer will indicate the relative humidity and beside it can be placed a thermometer.

If the specimen is particularly fine or of outstanding value and rarity, it really calls for a display case of its own. This may be constructed from heavy glass, laminated glass or bonded sheets of a clear plastic. It is advisable to have a substance such as silica gel, which will assist in keeping the air within the case dry, in a small drawer at the bottom of the case. Any special lighting that is used should have some form of diffuser fitted not only to provide a more pleasing effect but to avoid highlights that are too harsh.

If the collection is housed in an urban area or near some industrial complex, the atmosphere is likely to be polluted and there may be substances in the air which could cause tarnish and corrosion. As it is practically impossible to completely air condition a normal house in that, a filter system would be working and other safeguards, a decision will have to be made with regard to the use of some surface protectant if the objects are not all to be kept in cases. This does sometimes pose a difficult question as to which is the best course. Some lacquers are almost invisible when they have dried on the object; but they do need experience in applying, for, if the coating is uneven or small areas are missed, quite soon unsightly marks will appear, and if this occurs the lacquer will have to be completely stripped off and re-applied. The alternative to lacquering is the application of a very thin coat of a beeswax paste or a grease such as lanolin. With both these it is emphasized that the coating must be really thin, because if enough remains to leave the surface tacky, it will act as a trap for dust and grit particles.

The commonest metals and alloys to be met with will be iron, steel, copper, brass, bronze and silver. There may be rare instances of small amounts of gold but this metal is unlikely to pose any problems.

It is important to remember that, with all metals and alloys, it is very easy to go too far with any cleaning treatment and cause surface or other damage that will be costly and difficult to put right. The end result of any polishing or grime removal should be to complement the age of the object rather than to leave it with a fierce brash glister; a gleam perhaps, but never severe over-polishing that will almost certainly remove patina and will cause needless wear over engraved and chased decoration. All too often can be seen exquisite examples of old metal craftsmanship that have been seized by an ignorant hand and have been liberally soused with some harsh cleaner and then had the 'age' buffed right off them. Worse, some people bent on what they think is a fine polish often lean too heavily on a fragile specimen whilst holding it against a table or a hard edge, their idea being to get a purchase for the heavy handed rubbing they feel is necessary; the result can all too often be denting and scratching.

Selecting a polish can be difficult. There are so many tins, bottles, jars and aerosols on the supermarket's and ironmonger's shelves. Electrochemical cleaners of the 'dip' variety are intended to remove tarnish and, if the instructions on the container are followed, they should be safe, although other materials that may be next to the areas being treated should be masked off with an adhesive plastic tape, and finally the 'dip' should be rinsed off. The liquid, powder and paste polishes should have as fine a base to them as possible. It is a good idea to take a little between thumb and forefinger and rub it to and fro to judge just how gentle the abrasive substance in the polish is. There are a number of powders the manufacturer may use and these can include crocus, jeweller's rouge, pumice, putty, Tripoli and whiting; the crucial point is just how finely the powders have been ground. Certainly the best cleaning powder is jeweller's rouge. The presence of this in a proprietary polish can to a degree be judged by colour; the higher the percentage of rouge, the more distinctive the tint of warm red. For casual cleaning impregnated cloths are available, some of which may be safe to use on all metals and alloys, others may specify a particular purpose.

The components made from iron and steel will probably give the most trouble. Both of these in a damp or polluted atmosphere are very soon attacked. Once rust has taken a footing, it is very difficult to remove and is almost certain to leave behind some pitting. Should an instrument exhibit even the slightest sign of rusting, it must be dealt with as soon as possible. The simplest form of treatment is first of all to give the affected area a thorough soaking with paraffin to soften the rust. This can be done where practical by leaving small swabs of cotton wool soaked in paraffin over the affected places. After several hours the rust can be gently worked over with a gentle abrasive powder or very fine steel wool or a hog bristle brush. Great care will be needed to see that the removal is done evenly. Another course is to use a proprietary rust remover, and after this has soaked in for a period of time, as suggested on the instructions, the rust can be gently removed.

The complete dismantling of a scientific instrument by someone other than an expert is certainly not to be advised, although it may be possible to remove the base or some major part of the mounting safely. No matter how obstinate a joint or hinge is, no force should ever be used. One course is to inject a little penetrating oil, or even just paraffin will often suffice, to make the parts movable. Do not be tempted to use any acid on iron or steel, as it is very easy to cause damage that will be impossible to correct.

The treatment of copper, bronze and brass needs some thought. All three, with time, can acquire a patina which, particularly with copper and bronze, may be a valued sign of age. Also, this is not something that can be easily or effectively aped by the forger and should therefore be handled carefully during cleaning. Often, in the case of an instrument, there is likely to be a great deal of delicate engraving, either of figures, instructions or fine decorative features. It is important that this is protected as far as possible when cleaning. Any polish, whether paste or semi-liquid, should not be of a type that will readily clog fine lines or be a coarse scouring variety. Detailed work is best handled using a small hog bristle brush or a piece of cotton wool wound round the end of a wooden splint.

One of the proprietary electrochemical cleaners may be used with copper, but after use the areas must be thoroughly rinsed and dried. Silver inlays can be treated with one of these cleaners and it is best to apply them with the cotton wool and splint or a small soft-hair paint brush, again rinsing properly. One point to note is that a pot of electrochemical cleaner should only be used for one metal, and a fresh pot must be taken for each variety of metal treated. Small obstinate stains on copper or brass will sometimes yield to very weak nitric acid. This should not be of a greater strength than about 3 per cent and can be applied with a soft-hair paint brush or a piece of cotton wool on a splint. The application should only be for a matter of a few seconds and then the acid should be thoroughly rinsed away with distilled water. Rubber gloves should be worn and glasses for the eyes act as a safeguard against splashes. A second course can be to try a very gentle rubbing with a paste of olive oil and jeweller's rouge. Apply with a piece of cotton wool or a cotton bud and finish with a chamois.

Bronze, which is an alloy of copper and tin, should be treated with respect. This material does acquire, with age, a very fine patina which should not be disturbed by unwise cleaning methods. It is, sadly, very simple to remove this exquisite surface finish in a matter of minutes by the injudicious use of a too harsh metal polish. The soft warm brown or other tones then give way to a high polish which perhaps was not the intention of the original maker. The patina, if not deliberately applied by the craftsman, develops naturally through the action of the atmosphere over the years and, apart from being aesthetically pleasing, it helps to preserve the alloy. Gentle buffing with a soft chamois should suffice and it is best to avoid any stronger action, unless the bronze is already highly polished, when it can be treated in the same way as brass.

Another important point is with regard to lacquer that was probably applied by the original maker. This is likely to be of a shellac base and should as far as possible be preserved. It really should not be removed unless fully expert advice has been obtained, and any treatment should be left to a professional restorer.

It is possible that some instruments, particularly those of an ornate and elaborate style, may have areas decorated with a form of enamelling. This might be straight on to the metal or have been applied by the *champlevé* or *cloisonné* techniques. *Champlevé* is where the craftsman initially cuts small depressions or cells into the supporting metal and then puts enamel pastes into these before the whole is fired; the *cloisonné* technique means that thin metal wires are fixed to the base metal and become literally tiny fences in between the enamel pastes. The wires are usually of gold or silver, both giving a rich effect when the colour pastes have been fired. In general most enamels tend to exhibit a fairly rugged resistance to handling and the passing of time, but it is wise to inspect them at intervals. Use a strong glass and hold it so that the light falls at an angle across the piece. With a masterpiece, if there are signs of fresh cracking or any intimation of looseness, consult a trained restorer as soon as possible. Depending on the value of the object, there are two courses that can be followed if the piece is of minor importance. A clear lacquer can be sparingly applied so that it will sink down into the cracks or a thin synthetic water-clear adhesive can be applied in the same way.

An instrument may be found that is suffering from corrosion; the instrument may have been in a damp storing place, or the corrosion may have been encouraged by some atmospheric pollutant or, if by the seaside, long neglected exposure to the saline atmosphere will bring on an attack. Unfortunately, corrosion is basically a loss of metallic quality and any treatment should be left to the expert. There are now techniques available that can to a high degree ameliorate the condition. Professional help should also be sought for the removal of any dents, perhaps in the barrels of telescopes or microscopes, or severe scratching on the metal surfaces.

It is likely with the more ornate instruments to come across ormolu, brass or bronze gilded ornamentation. This can become dull and tarnished, because the underlying brass or bronze tends to 'sweat' through. This can be safely treated, but where possible the ornament should be dismounted from the main piece; if this cannot be done, adequate masking of the other materials should be made. One treatment can be to prepare a cleaner from 3 oz sodium potassium tartrate, 1 oz sodium hydroxide (caustic soda) and 2 pints distilled water. Shake these well together and then apply with a soft nylon brush. Leave the mixture on for a few minutes and then rinse thoroughly with plenty of clean water.

A second treatment is to mix together 1 part of alum to 30 parts of 3 per cent nitric acid. Apply as above with a soft nylon brush, only this time gently agitating for about a minute and then rinse well. With both these attempts it is wise to wear rubber gloves and be careful to avoid any splashing of the skin or eyes. If the tarnished appearance is only slight it can often be cleared by just sponging with a little warm soapy water to which has been added a spoonful of white vinegar. Do not be tempted to use metal polishes on ormolu as they will almost certainly take off the wafer-thin gold.

Engraving on metals, particularly with silver and sometimes with brass and others, should always be inspected closely before cleaning to see if there is any sign of the engraved lines having been 'filled' by the original craftsman rather in the way that silversmiths used the niello technique to fill in the engraved lines with a dark composition. If this is suspected, be extremely careful that it is not removed by excessive polishing with a brush: should this occur, consult a trained operator who will be able to undertake the replacement.

When working on the metal parts of an instrument it is a good idea to wear a pair of soft cotton gloves. It is very easy to scratch a finely worked metal surface with a fingernail or piece of hard skin on a fingertip. If using soft dusters or a chamois, examine them thoroughly to see that they do not contain or have not picked up any hard pieces or gritty dust.

Where leather has been used as barrel covering on telescopes or microscopes, it may well have started to perish if the instrument is fairly old. This should be treated as soon as possible with a dressing that is used by the

British Museum. It is quite simple to make up as long as adequate fire precautions are taken. The recipe consists of 7 oz anhydrous lanolin, $\frac{1}{2}$ oz beeswax, 1 fl oz cedarwood oil and 11 fl oz hexane. The hexane is extremely inflammable. Dissolve the beeswax in the hexane first and then add the lanolin and lastly the cedarwood oil; the result should be a thin cream. The dressing should be applied sparingly with a small piece of cotton wool and then the object left for two days. It will be found that the dressing will impregnate the old perished leather, tending to bring back the texture and fibre strength, and more than this, it will bring up the colours, the tooling and the gilding. After two days the surface of the leather can be gently polished with a piece of cotton wool, soft cloth or a soft-hair brush.

If the leather has started to peel away from the metal or wood on which it has been stuck, treat these areas first, before applying any leather dressing. As far as possible, remove all traces of the old adhesive. This can be done by using the small blade of a penknife or scalpel, taking care that the underlying metal or wood is not cut or scratched; a magnifying glass can be a help here. The best type of adhesive for refixing leather is one of the most modern synthetic resin emulsions, if possible one with a plasticizer, which is specifically intended for the materials it is being used on. Leather kept in overdamp atmospheres, particularly where there is inadequate ventilation, is liable to fungal attacks which can cause rotting, staining and colour changes in dyed skins. The mould and fungus problem can be treated in a number of ways. For a mild outbreak it may be sufficient just to spray the leather with a saturated solution of thymol in alcohol; take care to shield surrounding materials. If the trouble persists, it can be treated with a stronger fungicide, paranitrophenol, although this may cause some yellow staining. If it is diluted to a strength of 0·35 per cent this will be very slight. A third treatment is to use pentachlorophenol in a 0·25 per cent dilution.

Bone and ivory are both susceptible to damage by water and any treatment undertaken should involve the minimum of wetting. If there should be staining, the application of a paste made of whiting and 20 volume hydrogen peroxide is often successful. The paste should be made up to a stiff consistency and put on with a hog bristle brush or a small spatula. Rubber gloves should be worn to protect the hands, and as far as possible surrounding metal or wood surfaces should be masked. The operation should be carried out on a dry day and the paste should be left on, in the first instance, for about 15 minutes, then removed and the staining inspected. If further action is needed, the paste can then be re-applied and left on for another interval. When the desired state has been reached the area should be gently but thoroughly rinsed with small pieces of cotton wool damped with distilled water, and then dried with cotton wool or a soft cotton cloth.

To remove surface grime from bone or ivory, a 50–50 mixture of methylated spirits and distilled water is safest as it will evaporate more quickly. The liquid can be applied with a small piece of cotton wool or a cotton bud. If some of the dirt will not come away with this treatment, a small amount of powdered whiting can be mixed with the distilled water and methylated spirits. After cleaning or stain removing, the bone and ivory parts can be given a gentle application of a little almond oil rubbed in with a piece of cotton wool. This will bring up the original sheen of the material.

Where wood has been used for the mounts, the base or other parts, it should first be carefully inspected for any signs of attack from woodworm. If there is baize underneath the base, examine this also for traces of egg laying or damage from moths and larvae. If there is any evidence of woodworm, treat the holes with one of the proprietary liquids. One convenient way is to use a pressurized aerosol container, as this has a fine nozzle which will inject right into the holes. Always wear glasses or goggles to protect the eyes because, very often, the poisonous liquid being injected can suddenly squirt out of a nearby hole. After treatment, surplus liquid should be wiped away from the surfaces and the wood left for several days to allow the toxic vapours to evaporate.

Old woodworm holes and cracks may be filled with beeswax that has had some powder colour mixed with it to match the tone and tint of the wood. For small crevices this method will give a more satisfactory after-appearance than plastic wood or other fillers. If parts of the wood have become seriously weakened by woodworm attack or have rotted, a decision will have to be taken as to the next step. As far as possible, it is highly desirable to retain the original material. Modern synthetic and bonding agents are available which can work wonders with wood that has reached a state of serious perishing. Impregnation with these materials, however, is not a procedure to be carried out in the home and the advice of an expert should be sought.

The best surface treatment for wood is with beeswax polish. There are proprietary brands of this that can be bought, but it is quite simple to make up. Melt some yellow beeswax in an old saucepan and when it is completely liquid, remove from the heat source and carefully stir in a little pure turpentine. By adding more or less turpentine the consistency of the final paste can be made as stiff or creamy as desired. Always apply the polish sparingly and leave it on for a few minutes, then buff up with a soft cloth or piece of cotton wool.

Glass dials or other parts, if grimy, can be cleansed with a 5 per cent solution of ammonia in distilled water. This may be applied either with cotton wool on a wooden splint or a small soft-hair paint brush. Afterwards the areas should be thoroughly rinsed and dried. If there are any pieces made of porcelain which have painted or inscribed words or figures on them they should be treated with circumspection, as any liquid treatment could take off the inscriptions.

The lubrication of any moving parts should only be done using one or other of the non-drying oils and with the minimum application. The oil chosen should be clear and not too viscous; one of the most suitable is porpoise oil.

FORGERY AND FAKES

As with other sought-after collector's items, scientific instruments have drawn the attention of the skilled but crooked craftsman. Most genuine items by the nature of their creation must be in strictly limited supply so if there comes a surge in collecting the forger sees to it that the demands of a ready market are met. Forged scientific instruments produced with an understanding and consummate skill are not all that common although there are quite a number of rough ignorant examples which are unlikely to deceive a specialist.

A problem also arises when a fake does get through the barriers of scholarship and examination and ends up as a prize piece in some collection. The connoisseur or dealer can be averse to admitting that he or she has been taken in. Sadly it is not unknown for a fake, once acquired, to remain concealed and as time passes the years bring a false sense of provenance and quality. Another factor on the side of the forger is that if he is really skilled and has done his research well the first batch of his deceptive goods stands a fair chance of getting through and may even pass for a number of years before the perceiving eye picks them up. A maxim when acquiring is never to buy in haste, take time to research the object, look well and long, perhaps go away and come back the next day and go on looking. If the object has been faked this concerted observation can very often pierce the cleverness of the forger, expose minute mistakes of finish, design or even stupid anachronisms. What at first encounter may have given the impression of being an object of skill and beauty may start to fade and can be seen for what it is – a nasty attempt at dishonesty and worse, one that if it enters the channels of collecting can not only cause loss financially but also upset the whole fabric of a rather wonderful history of discovery and accomplishment.

The story of forging scientific instruments starts somewhere around the end of the 17th century. Collectors became engrossed with the advances of science and it became the 'in thing' to add these items of metal, glass and other materials to the collector's cabinet. Probably those investing in such things felt themselves partially at one with the active minds of the great men who were breaking through the barriers of ignorance.

Some of the efforts of the forger seem hardly to be worth while, but he is in all likelihood mass-producing them and through a carefully worked out marketing system can spread false merchandise around the world. His outlets can include small-time sale rooms, back street dealers, house sales, even the hoary attempt to pass something to someone in a bar. Towards the end of the last century, collectors in Europe were plagued by a sudden wave of water clocks that appeared with a rush on the market – they were being hacked out in the hidden sweatshops in Birmingham. Another item often faked is the early Nuremberg weight. From time to time sets of these surface, often so badly made that they will not nest properly and perhaps even have wrong markings and inscriptions.

The main areas for the collector of scientific instruments are items concerned with astronomy, navigation and measurement. Of these it is astronomy that seems to exert the strongest lure; the sense of the great endless space can engender a feeling of romance and wonder, that can, if not watched for, obscure the judgement. One of the key instruments for the early astronomer was the astrolabe. With this he could take essential first steps for observation; the astrolabe would give the altitude of the sun, stars and moon. Many astrolabes are comparatively simple and comprise a series of flat metal plates, often beautifully figured, lettered and decorated, which revolve within the same diameter; some are pierced to allow the observer to register several measurements at the same instant. The key plate, which is called the 'rete', is a star chart as seen by a viewer looking back at the known universe from outer space. The finest of these instruments are highly accurate and embody an advanced knowledge of astronomy. It can often be this factor that will give the deceiver away, coupled also with the simple fact that the exquisite object just does not work as it should.

In March 1976 a Swiss dealer bought at auction a brass astrolabe for £3,400. Somehow rumours started that it could be a fake and the Art and Antiques Squad from Scotland Yard were called in to investigate. At the same time Sotheby Parke Bernet issued a statement which caused a considerable tremble to vibrate though the trade and collectors' world. Sotheby's intention was to reassure its own clients but it did at the same time raise a general alarm over such matters. The document read:

> From time to time in the history of the art market a master forger has emerged. Amongst them have been Bastianini, Van Meegeren and Thomas Wise. It is alleged that a forger of scientific instruments may have successfully hoodwinked collectors, dealers and auctioneers. Sotheby's, as the only auctioneers in Great Britain who provide a five year guarantee against forgery, naturally stand behind their conditions of sale and will not only make restitution to any purchaser who has in fact been damaged as a result of the alleged fraud but in this particular case, will also make restitution to such purchasers of instruments from this source purchased

before the implementation of this five year guarantee period, provided that forgery is established to Sotheby's complete satisfaction.

One scrap of evidence of an earlier date turned up. The French magazine *Art et Curiosité* of September 1974 carried an article by a leading authority on astrolabes – the Parisian dealer Alain Brieux. In this article he mentioned that he had been caught with a false instrument and it was possible that it had been made by the same English forger that Scotland Yard was now trying to trace. One of Monsieur Brieux's regular customers for books about old scientific instruments had called with another Englishman who had sold him what he had taken to be a genuine 18th-century ring dial – admittedly an unusual one as it combined an astrolabe and an astronomical ring. Other French dealers reported that they too had been buying similar objects. Monsieur Brieux set out to make a careful examination of a number of these unusual ring dials. One fact was that if they were truly of the date stated they would have been constructed of brass beaten to a near uniform thickness by hand. When an accurate analysis of the metal thickness was made it became obvious that the sheets had not been hand beaten as the thickness was always exactly the same. The brass had to be modern. With this indisputable clue as a spur the examination continued and another slip was uncovered. The dials were signed by various French and German makers but the trained eye found a stylistic similarity which pointed to the fact that the instruments just could not have been made by different people. The forger perhaps carried away with his own prowess added small engraving conceits to the letter 'S' and also the figure '5' on each of the by now discredited instruments. It is with such apparently small details that the faker can give himself away – prove one point of falsity and others will appear. The villain in this case was certainly a trier. He offered Monsieur Brieux what he claimed was an extreme rarity. The object posed as a nautical hemisphere, so rare in fact that there is no record of one ever having been actually made, but there is an illustration of a design for one in a book by Michel Coignet which the faker must have seen. He had not realized that the item had never left a craftsman's shop. He asked for 20,000 French francs, which was refused and another step had been taken towards his arrest. The law and the process of justice removed this gentleman from circulation. His secret workshop in Slough, Buckinghamshire, was closed down. In his way he epitomized many of the skills of forgers in this line of business. He had simulated the system of 'moss-scratches' which a metal will acquire over the years of handling. This can be done by placing the pieces of the instrument being made in a box with sundry scraps of jagged metal, old bolts and the like and then giving the box a lengthy shaking about. A little time spent with weak nitric acid can soon produce creditable signs of age; old treatises will give other, more disreputable instructions for weathering or patinating metals to give them the required appearance of age. Instructions can also be found for preparing 18th-century lacquer finishes. Notes give secrets for treating the leather sometimes used on the barrels of microscopes and telescopes; ivory and bone details can all be treated to give advanced age as can wooden mounts or cases.

There is one area for particular care and that is with reproductions. These are actually rare and would probably have been in the first instance produced with a totally honest intent. One such case was the habit of some American pharmaceutical companies of giving excellent replicas of such instruments as microscopes to respected and otherwise largely incorruptible members of the medical profession. Japanese doctors in particular prized such items for display. One example of this was a reproduction of an early microscope designed by John Cuff, complete with wooden case, ivory study slides, fish carrier and a carefully printed facsimile of a Cuff label. The modern maker, however, had built into the instrument a safeguard for the collector. In the support arm of the microscope was a small steel magnet. As the instrument itself was made entirely of brass anybody with a small pocket compass who knew where to hold it could not be deceived, although it would take a little nerve to start examining such items in a sale room or dealer's salon. Nevertheless, the nagging doubt remains with such perfection of craftsmanship that an hour or two with nitric acid, caustic soda, false lacquer, scratching and general mishandling could produce a very saleworthy item.

TABLE OF EVENTS

A chronological summary of the most significant landmarks in the history of scientific instruments

BC

10000–5000 Scratched representations of stars on rock.

3000–2500 Erection of the pyramids at Gizeh, Egypt. Orientation to points of the compass and exactitude of construction point to the fact that the builders would have had some form of precise instruments.

3000–2500 Chinese determining equinoxes and solstices.

2100 Babylonians' calendar changed to be in accord with the movement of the stars.

585 Thales of Miletus predicts an eclipse.

*c.*200 Eratosthenes probably uses a form of solstitial armillary sphere and an astrolabe. He calculates the circumference of the earth.

200 Atlante Farnese celestial globe.

*c.*135 Ctesibius of Alexandria develops a clepsydra.

*c.*100 Seneca uses globe of water to magnify writing.

AD

*c.*140 Ptolemy produces his *Almagest.*

*c.*1000 Ibn Junis compiles Hakimite Tables.

1038 Arab philosopher Alhazen investigating idea of a camera obscura.

1252 Alfonsine Tables prepared under the patronage of Alfonso X.

*c.*1260 Hulagu Khan establishes Observatory at Maragha.

1288 Balance clock on former clock tower at Westminster.

1348 Astrarium made by Giovanni de'Dondi.

1420 Ulugh Beg builds Observatory at Samarkand.

1498 John Cabot lands in Greenland.

1530 Gemma Frisius writes a treatise on astronomy and cosmogony.

1541 Mercator produces terrestrial globe.

1543 Copernicus's *De revolutionibus orbium coelestium* published.

1561 Landgrave of Cassel builds first observatory with revolving dome.

1580 Tycho Brahe completes Observatory at Uraniborg.

1582 Gregorian Calendar replaces the Julian.

1594 Napier begins research leading to discovery of logarithms

1597 Danfrie publishes description of a graphometer.

1600 Praetonius describes a mensula.

1608 Lippershey discovers power of two lenses to magnify distant scenes.

1609 Galileo makes his first telescope.

1611 John Speed brings out *Theatre of the Empire of Great Britain.*

1627 Kepler prepares Rudolphine Tables.

1642 Pascal invents calculating machine.

1643 Torricelli discovers principle of barometer.

*c.*1650 Father Chérubin constructs binocular microscope.

1659 Boyle and Hooke construct 'pneumatical engine'.

*c.*1660 Bloud making diptych dials.

1663 Gregory proposes reflecting telescope.

1668 Newton's reflecting telescope.

1675 Original Royal Observatory at Greenwich designed by Sir Christopher Wren.

1675 Huygens makes spiral spring for balance wheel.

1679 Hevelius observatory, books and instruments destroyed by fire.

1680 Quare develops repeating watch.

1696 Earliest mention of a metronome.

1700 John Marshall constructs early compound microscope.

1713	John Rowley constructs original orrery.
1720	Simson produces true theodolite.
1725	Catalogue of stars by Flamsteed published.
1730	John Hadley makes a sextant.
1731	John Hadley constructs successful reflector telescope.
1731	Réaumur invents his thermometer.
1742	Wilhelm Jacob discusses design for a heliostat.
1746	George Adams makes his 'New Universal Microscope'.
1748	Joseph Bramah patents his lock.
1758	Mudge invents lever movement for clocks.
1761	Harrison produces accurate chronometer.
1767	First *Nautical Almanac* brought out.
1776	Herschel constructs 7 ft telescope.
1781	Uranus discovered by Herschel.
1783	Jacques Alexander César Charles is first to send up hydrogen filled balloon.
1783	Herschel publishes his *Notions of the Solar System*.
1785	Bramah invents hydraulic press.
1787–8	Herschel constructs 20 ft telescope.
1789	Herschel constructs 40 ft focal length telescope with 4 ft mirror.
1789	Ramsden makes 5 ft vertical circle for Piazzi.
1793	John Dalton publishes his *Meteorological Observations and Essays*.
1795	Admiralty Hydrographic Office opened.
1814	Fraunhofer invents heliometer, discovers lines of solar spectrum, prepares way for spectroscopy.
1817	Brewster rediscovers kaleidoscope.
1821	Faraday begins research on electromagnetism.
1827	Ohm formulates his law.
1839	Nasmyth invents steam hammer.
1840	Dr Robinson invents the anemometer.
1844	Robert Hunt develops ferrotype.
1845	Earl of Rosse completes 6 ft reflector telescope at Birr.
1855	Bunsen invents his burner.
1856	Henry Fitz constructs $12\frac{1}{2}$ in refractor telescope.
1863	Angelo Secchi announces his system for classifying spectra of stars.
1876	Graham Bell transmits sound by electricity.
1877	Schiaparelli observes canals on surface of Mars.
1895	Röntgen discovers rays, leading to X-rays.
1898	Yerkes refracting telescope erected, largest in the world.
1901	Marconi sends wireless signal from Poldhu, Cornwall to St John's, Newfoundland.

INSTRUMENTS THROUGHOUT THE WORLD

Selected places where examples of scientific instruments may be seen

AUSTRALIA

Montville	Elands Old Clock Display and Mini-Museum
Sydney	Macleay Museum
West Gosford	Kendall Cottage

AUSTRIA

Arnfels	Museum of Clocks
Bad Ischil	Museum of the Haenel and Pancera Families
Feldkirch	Heimatmuseum
Graz	Joanneum Provincial Museum of Cultural History and Applied Art
Klosterneuburg	Museum of the Augustinian Monastery
Kremsmünster	Monastery Collections
Langenlois	Regional Museum
Oberperfuss	Anich Hueber Museum
Pöls-Enzersdorf	Husslik Regional Museum
Reutte	Regional Museum
Vienna	Clock Museum; Kunsthistorisches Museum; Museum of Industry and Technology; Sobek Collection; Treasury of the German Order

BELGIUM

Antwerp	Provincial Museum of Industrial Art
Brussels	Musée Royaux d'Art et d'Histoire
Ghent	Museum of Antiquities; Museum of the History of Science
Liège	Musée de la Vie Vallonne
Lier	Astronomical Clock
Saint-Trond	Béguinage Museum

BRAZIL

Goiás	Museum of the Bondeiras

BULGARIA

Sofia	National Polytechnic Museum

CANADA

Edmonton	Queen Elizabeth Planetarium

CHINA PR

Peking	Imperial Palace; Old Observatory; Planetarium
Tiensin	People's Hall of Science

CZECHOSLOVAKIA

Bratislava	Municipal Museum
Košice	Technical Museum
Prague	National Technical Museum
Přerov	J. A. Komensky Museum

FINLAND

Espoo	Clock Museum
Helsinki	Museum of Technology
Turku	Maritime Museum

FRANCE

Beauvais	Astronomical Clock
Besançon	Museum of Clocks and Watches
Bourges	Hôtel Lallemant
Clermont-Ferrand	Musée de Panquet
Molsheim	Museum of Art and History
Mont-Saint-Michel	Museum of History
Paris	Astronomical Museum of the Paris Observatory; National Museum of Arts and Crafts; National Technical Museum
Strasbourg	Museum of Decorative Arts
Toulouse	Paul Dupuy Collection

GERMANY, EAST

Dresden	Martin Anderson

	Collection; State Hall of Mathematics and Physics
Meissen	Municipal District Museum
Potsdam	Bruno and Burgel Collections

GERMANY, WEST

Bamberg	Historical Museum
Barmstedt	Museum der Grafschaft Rantzau
Flensburg	Städtisches Museum
Friedberg	Regional Museum
Furtwangen	Historic Clock Collection
Jever	Coastal Museum
Michelstadt	Odenwald Museum
Nuremberg	Germanisches National Museum
Munich	Bavarian National Museum
Neustadt	Regional Rooms
Rendsburg	Schleswig-Elektro-Museum
Schweinfurt	Städtisches Museum
Schwenningen	Hellmut Keinzle Clock Museum; Schwenningen Regional Museum
Stuttgart	Württemberg Museum
Wuppertal	Friedrich-Engels-Gedenk-statte; Wuppertal Clock Museum

GREAT BRITAIN

Basingstoke	Willis Museum
Birmingham	Avery Historical Museum; Museum of Science and Industry
Bury St Edmunds	Gersham-Parkington Memorial Collection of Clocks and Watches
Cambridge	Whipple Museum of the History of Science
Dundee	City Museum and Art Gallery
Glasgow	Kelvingrove Museum
Leicester	Newark House Museum
Lincoln	Usher Gallery
Liverpool	City Museum

London	British Museum; Clockmakers Company Museum; Dollond and Aitchison Ltd; National Maritime Museum; Old Royal Observatory; Science Museum; The Queen's Gallery
Manchester	Museum of Science and Technology
Norwich	Bridewell Museum
Oxford	Museum of History and Science
Redcar	Museum of Shipping and Fishing
Rotherham	Corporation Department of Trading Standards
Tiverton	Tiverton Castle
Winchester	The Westgate Museum

HUNGARY

Sárospatak	Museum of the Reformed Church

INDIA

Calcutta	Birla Industrial and Technical Museum
Dhulia	I.V.R. Rajwade Lanshodhan Mandal Museum
New Delhi	Archeology Museum

IRELAND

Birr	Birr Castle Museum
Dublin	Egestorff Collection

ISRAEL

Haifa	National Maritime Museum

ITALY

Bologna	Museum of the Institute of Astronomy
Florence	Cartography and Instruments Museum; National Museum for the History of Science
Milan	National Museum of

	Science and Technology – Leonardo da Vinci
Rome	Astronomical and Copernicus Museum; Collection of the Astronomical Observatory
Treviso	Regional Museum
Trieste	Civico Museo del Mare

JAPAN

Ikoma-Gun (Nara Prefecture)	Museum of Universal Science
Kyoto	Board of Education Science Room
N'igata	Science and Technology Museum
Tokyo	Museum of the National Research Laboratory of Metrology; National Science Museum

NETHERLANDS

Bussum	E. A. Veltman Collection
Delft	Electro-Technology Study Collection
's-Gravenhage	Museum of Weights and Measures
Gröningen	University Museum
Haarlem	Teylers Museum
Leyden	National Museum of the History of the Natural Sciences
Rotterdam	Historical Museum
Utrecht	Gold, Silver and Clock Museum

NEW ZEALAND

Whangarei	Clapham Clock Collection

POLAND

Frombork	Copernicus Museum
Kraków	University Museum of Art
Rozewie	Museum of Lighthouses
Toruń	Copernicus Museum
Warsaw	Meteorological Museum
Wloclawek	Jozef Arentowicz

	Collection of Measuring Instruments

PORTUGAL

Lisbon	Museu Etnografico do Ultramar

RUMANIA

Ploesta	Museum of Clocks

SPAIN

Coruña	Clock Museum
Madrid	Museo Naval

SWEDEN

Stockholm	Technical Museum

SWITZERLAND

Altdorf	Uri Historical Museum
Basel	Kirschgarten
La Chaux-de-Fonds	Watch and Clock Museum
Delémont	Jura Museum
Gerewon	Museum of History and Science
Winterthur	Jakob Bryner and Konrad Kellenberger Collections

USA

Chicago	Adler Planetarium and Astronomical Museum
Columbia	Museum of Horology
Flagstaff	Museum of Astrogeology, Great Meteor Crater
Lexington	Transylvania Museum
Los Alamos	Bradbury Science Hall
Los Angeles	Griffith Observatory and Planetarium

USSR

Leningrad	The Hermitage
Moscow	Museum of the History of Microscopy; Sternberg Institute

YUGOSLAVIA

Belgrade	Museum of Applied Art

PRICE GUIDE

Specimen prices paid for scientific instruments in salerooms, March 1983

£49 $78 Brass spectroscope. English, *c.*1920

£85 $136 19th-century brass circular protractor by Cail, Newcastle

£90 $144 Brass pantograph by A. Abraham & Co., Liverpool

£90 $144 Brass inspector's beam scale by De Grave & Co.

£90 $144 Spy-glass in tortoiseshell and ivory. G. Adams, London

£110 $176 Oxidized brass surveyor's level by John Pardry, Durban

£120 $192 Portable balance with steel beam and brass pans

£143 $229 Late 19th-century Stanley brass box sextant

£160 $256 Stereoscopic microscope by Vickers Instruments, York

£176 $282 Necessaire complete with dividers, compass, pen, ivory protractor and sector. Signed T. T. Blunt, London

£180 $288 Surveyor's level by Troughton & Sims, 1844

£190 $304 19th-century brass tacheometer by Cooke, Troughton & Simms, London

£200 $320 Ebony octant by Troughton & Simms

£200 $320 19th-century brass chondrometer by Young & Sons, Bear Street, London

£210 $336 Barograph by Chadburn's Ltd

£210 $336 19th-century brass Lords patent calculator by W. Wilson, London

£220 $352 Sextant by W. Ludolph, Bremerhaven

£220 $352 Surveyor's level by F. W. Breithaupt & Sohn, Cassel

£240 $384 Brass miner's dial by J. Davis & Son, London and Derby

£260 $415 Brass ship's log

£280 $450 19th-century octant by T. Helmsley & Son, Tower Hill, London

£297 $475 Mid 19th-century dry card binnacle compass

£310 $496 Brass solar microscope

£320 $512 Mid 19th-century celestial globe by Kirkwood

£340 $544 Mid 19th-century brass astronomical telescope by W. & S. Jones, London

£370 $592 Brass microscope of Culpeper design

£380 $608 Late 17th-century silver perpetual calendar

£380 $608 Oxidized brass theodolite, gimbal mounted by Stanley

£380 $608 19th-century brass compound binocular microscope by Baker, Holborn, London

£385 $616 English brass lighthouse clock

£385 $616 19th-century Indian ivory quadrant

£418 $670 Early 18th-century brass universal equatorial dial, initialled LTM, Augsburg

£420 $672 3 in reflecting telescope by Adie, London

£440 $704 19th-century 3 in brass refracting telescope by Dollond, London

£440 $704 18th-century brass equinoctial dial signed Baradelle, Paris

£500 $880 Mahogany waywiser signed Cole, Fleet Street, London

£500 $800 Transit theodolite by Troughton & Simms, London

£500 $800 18th-century brass pedometer

£520 $832 Mahogany wheel clock barometer by J. Flora, Nottingham

£540 $864 18th-century ebony and brass mounted quadrant by John Goater

£550 $880 Combined theodolite and miner's dial by Troughton & Simms

£650 $1040 30 in terrestrial globe by W. & A. K. Johnston Ltd

£660 $1056 18th-century brass reflecting telescope signed Fran' Watkins, Charing Cross, London

£660 $1056 Early 19th-century 3 in Harris pocket globe

£700	$1120	Universal ring dial by Troughton & Simms
£715	$1144	6 in Stanley brass transit theodolite, *c.*1900
£720	$1152	Compound monocular microscope by Dollond, London
£750	$1200	18th-century French brass graphometer by Clerget, Paris
£750	$1200	18th-century pasteboard and brass mounted four-draw refracting telescope by I. Cuff, London
£825	$1320	Mid 18th-century brass sector signed I. Finch Fecit
£850	$1360	Late 17th-century French Butterfield dial, signed Butterfield, Paris
£1050	$1680	17th-century German perpetual calendar combined with aide-mémoire, signed M. Mettlin Fecit
£1100	$1760	Gregorian 5 in reflecting table telescope by D. Jones, Charing Cross, London
£1100	$1760	Troughton brass double frame sextant, 1815/16
£1150	$1890	18th-century brass universal equinoctial dial
£1200	$1920	18th-century mahogany back staff
£1200	$1920	Brass ring dial, signed Richard Abbott Fecit (Richard Abbott was apprenticed to Helkiah Bedford, 1668)
£1210	$1936	Early 17th-century ivory universal altitude dial, possibly French
£1300	$2080	Lacquered brass microscope by Dollond, London
£1350	$2160	18th-century silver inclining dial by Heath & Wing, London
£1350	$2160	18th-century Dutch brass sextant, signed G. Hulst van Keulen Fecit, Amsterdam
£1400	$2240	Chest microscope by Nairne, London
£1400	$2240	Early 19th-century orrery by W. Jones, London
£1700	$2720	Universal pocket sundial by Norry, Gisors, 1644
£1800	$2880	18th-century brass circumferentor, signed Thomas Heath, London, Fecit
£1850	$2960	Ivory diptych dial dated 1650, with mark of Leonhardt Mire of Nuremberg
£1980	$3168	Late 18th-century brass compound microscope, signed Geo. Adams
£3000	$4800	18th-century brass altazimuth theodolite, signed Heath & Wing
£3300	$5280	Mid 19th-century brass kaleidoscope
£3800	$6080	Oak stick barometer in the manner of John Patrick
£10500	$16800	Mid 17th-century brass Maghribî astrolabe

FRAGONARD

Jean-Honoré FRAGONARD
Self-portrait
PARIS, Fondation Custodia (F. Lugt collection), Institut Néerlandais. Roundel in black chalk
17·5 cm. Annotated by Fragonard: se ipsum delineabat frago/apud de Bergeret/anno 1789.

FRAGONARD

DAVID WAKEFIELD

ORESKO BOOKS LTD·LONDON

ACKNOWLEDGEMENTS

I would like to express my gratitude to the following people who have helped me in various ways: Dr. Anita Brookner of the Courtauld Institute, who first awakened my interest in the French eighteenth century; Professor Francis Haskell of Oxford University for generously allowing me to make use of some of his ideas, particularly the substance of a lecture given to the Taylor Institution, Oxford, in 1975; Professor Jean Seznec of All Souls College, Oxford, for his constant willingness to answer my queries; the staff of the Witt Library; and Christopher Wright for help of many different kinds, but above all for his constant enthusiasm and interest in this project.

I would also like to thank all the following for their help in providing photographs: Banque de France, Paris; Bowes Museum, Barnard Castle; British Museum, London; Calouste Gulbenkian Foundation, Lisbon; Dr. Emile Bührle, Zürich; Cailleux Collection, Paris; Detroit Institute of Arts, Detroit; Ecole des Beaux-Arts, Paris; Frick Collection, New York; Fitzwilliam Museum, Cambridge; Mrs. Fowler Mc-Cormick, Chicago; Hermitage, Leningrad; Institut Néerlandais, Paris; Los Angeles County Museum of Art, Los Angeles; Metropolitan Museum of Art, New York; Museum of Art, São Paulo; Musée des Beaux-Arts, Besançon; Musée des Beaux-Arts, Rouen; Musée des Beaux-Arts, Tours; Musée Cognacq-Jay, Paris; Musée Fragonard, Grasse; Musée du Louvre, Paris; Museum of Modern Art, Barcelona; Musée de Picardie, Amiens; National Galleries of Scotland; National Gallery of Art, Washington; The Rosenbach Foundation, Philadelphia; Sterling and Francine Clark Art Institute, Williamstown; Stichting Collectie Thyssen-Bornemisza, Amsterdam; Toledo Museum of Art, Toledo, Ohio; M. Arthur Veil-Picard, Paris; Wallace Collection, London; Worcester Art Museum, Worcester, Massachusetts; Photographie Bulloz; Photographie Giraudon; Service de Documentation Photographique de la Réunion des Musées Nationaux; John R. Freeman & Co. Ltd.; and Mrs. Sabine McCormack and Ms. Peggy Edwards of Phaidon Press.

First published in Great Britain by
Oresko Books Ltd., 30 Notting Hill Gate, London W11

ISBN 0 905368 01 0 (cloth)
ISBN 0 905368 05 3 (paper)
Copyright © Oresko Books Ltd. 1976

Printed in Great Britain by
Burgess and Son (Abingdon) Ltd., Abingdon, Oxfordshire

Jean-Honoré Fragonard

THE GONCOURT BROTHERS, whose *L'Art du dix-huitième siècle* must still rank as the best general account of French eighteenth-century art, began their 1865 study of Fragonard with the statement that, after Watteau, he was the only true poet of his time. They went on to define Fragonard as 'the uninhibited raconteur, the gallant amoroso, pagan, playful, with Gallic malice, an almost Italian temperament and a French mind'. This lyrical, poetic strain which the Goncourts perceptively divined in Fragonard, despite their incomplete knowledge of his work, seems to have escaped many commentators, and to this day, in English-speaking countries at any rate, he remains one of the least known and least understood French artists. To the layman his name is familiar only as the author of erotic subjects, nubile young girls and their impatient lovers, all with an obvious appeal. This was the aspect of Fragonard which made him popular and successful in his own day with the rich financiers, actresses and demi-mondaines, who probably saw in him no more than a very accomplished pornographer, good enough for dashing off a spicy scene for their private apartments and for the amusement of their friends. In fact, erotic subjects count for only a fairly small part of Fragonard's total output, and it is doubtful whether they are his best and most original paintings.

Even today Fragonard has retained this aura of the connoisseur's painter, frivolous, light-hearted, technically brilliant but essentially vacuous, in short, typical of the decadent art of the Ancien Régime which earned the well-deserved strictures of Rousseau, Diderot and later Romantic critics. It is hoped that this study will show this view of the painter to be both misleading and incomplete. Like all prolific artists Fragonard can sometimes be trivial and banal, but at his best he is a great artist and deserves to rank with Watteau or Tiepolo, the master of a range of subject and mood unrivalled in eighteenth-century France. He possessed not only technical virtuosity, but also a powerful creative imagination which frequently bordered on the epic and fantastic, linking him retrospectively with the names of Tasso and Ariosto. The extraordinary vitality and sometimes visionary quality of his work carry us far beyond the confines of the Rococo world he is supposed to incarnate.

Jean-Honoré Fragonard was born in Grasse near Aix-en-Provence on 5 April 1732. He was the son of a glove maker who speculated his entire fortune in a disastrous enterprise, probably in fire engines, and in 1738 came to Paris to try to retrieve his capital outlay. This Provençal background, with its strong light, vivid colours, cypress trees and olive-groves, evoked in rhapsodic terms by the Goncourt brothers in their essay, left an indelible stamp on Fragonard's mind, for he remains always, unmistakably, a Mediterranean artist. The importance of milieu, however, should not be exaggerated, for Fragonard later showed that he was equally sensitive to the northern Dutch landscape, and this Protean adaptability and readiness to take advantage of any new opportunity is in itself typical of the easygoing southern temperament. Fragonard is one of the least consistent of all artists. In Paris, where he followed his parents, he was first apprenticed as a clerk to a notary, who, spotting the boy's aptitude for drawing, urged Fragonard's mother to send him to a painter's studio. In 1747 the boy began his studies under Boucher, Mme. de Pompadour's chief protégé, then at the height of his fame. Boucher, however, had no time for a completely untrained pupil and sent him on to Chardin. Fragonard was no more successful under Chardin, and after about six months the master made it clear to his pupil that he was not satisfied with his progress. We can well imagine that Chardin's advice, 'You seek, you scrape, you rub, you glaze, and when you have got hold of something that pleases, the whole picture is finished', and his laborious teaching methods based on copying prints were much too slow for someone of Fragonard's impetuous temper. Meanwhile, in his spare time, Fragonard copied the paintings he saw in the Paris churches and private collections. It was perhaps on the strength of these copies that around 1748 Boucher decided to take Fragonard back into his studio, where he spent the next four years.

Boucher's influence is predominant in Fragonard's early works like the *Woman Gathering Grapes* (Plate 1), *The See-saw* (Plate 4) and *Blind Man's Buff* (Plate 3), with their light tonality of grey-blue, green and pink and with the contrived disorder of their setting, what the Goncourts called 'le fouillis' and what Marie Antoinette created in real life at the Petit Hameau at

Versailles. Fragonard never entirely abandoned Boucher's preference for the light palette, but he far exceeded his master in imaginative power when he later transformed Boucher's anodyne pastorals into such overwhelming creations as *The Fête at Saint-Cloud* and *The Fête at Rambouillet*. For Boucher nature was merely a picturesque backdrop in which plants and shrubs could be shifted at will like theatrical props. Theatre and painting, in fact, were closely linked throughout the eighteenth century, and Boucher frequently designed scenery for such heroic pastorals as *Issé*, by Destouches and Lamothe. Critics saw him as the natural heir to the tradition of the gallant eclogue, stretching from Virgil to Fontenelle, but after the middle of the century the public began to tire of the false gentility of Boucher's marquises dressed up as shepherdesses and looked for something more authentically rustic and with stronger human feeling. The difference between Boucher and Fragonard can be measured if we compare their respective treatment of two subjects from Tasso, Boucher's *Sylvia and the Wolf* (fig. 1) of 1756, painted for the Hôtel de Toulouse, with Fragonard's slightly later *Rinaldo in the Gardens of Armida* (c. 1765) (Plate 13). Boucher's picture is pretty and charming, but passive, totally undramatic, whereas Fragonard's is all violent movement, the hero in his plumed helmet dashes on the scene, Armida sweeps down to greet him, and the trees and clouds are caught up in a vortex of colour. Between 1750 and 1765 Fragonard transformed Boucher's technique out of recognition, and it is ironic that Boucher's parting advice to Fragonard, on the departure of the younger man from his studio, was not to take the Old Masters seriously.

In 1752 Boucher urged Fragonard to compete for the Prix de Rome, even though he had received no formal academic training. He submitted for the competition *Jeroboam Sacrificing to the Golden Calf* (Plate 5), a competent work in the required historical style which he mastered with surprising ease and promptly won the prize. Before going on to Rome Fragonard had to spend four years at the Ecole Royale des Elèves Protégés, an academic school installed in the courtyard of the Louvre, under the directorship of Carle Van Loo. The curriculum consisted of drawing from life and from statues, reading from Bossuet and the classical authors and making copies after Old Masters from Paris private collections and neighbouring châteaux. In 1756 Fragonard went on to the Ecole de Rome, the mecca of any aspiring young artist in the eighteenth century, and it was probably there, in 1757, that he met the young Greuze. The director at that time was the popular and easy-going Natoire, who kept up a regular correspondence with the marquis de Marigny, Directeur des Bâtiments in Paris and the brother of Mme. de Pompadour. These letters give us an idea of Fragonard's activities in Rome, where, despite the climate of tolerant

fig. 1 François BOUCHER
Sylvia and the Wolf
Oil on canvas 122·5 × 133 cm. 1756. Tours,
Musée des Beaux-Arts

One of a series of four oval overdoors illustrating Tasso's poem *Aminta*, painted for the duc de Penthièvre's Paris town house, the Hôtel de Toulouse, now the Banque de France. Two of the paintings are still there, the other two are in the Musée des Beaux-Arts, Tours.

paternalism, he does not seem to have been entirely happy. At that time there was a growing conflict between academic teaching and the new, fashionable ideas favouring freedom and spontaneity; Fragonard was caught in the cross-fire. Nor were the authorities themselves in agreement about the best training for an artist. Marigny seems to have insisted on strict, regular copying from Antiquity and the Old Masters, while Natoire was more inclined 'to allow genius a little freedom' (30 August 1758), fearing that 'the imitation of certain masters may do him harm' (31 August 1758). In October, we learn, Fragonard was making progress on a copy after Pietro da Cortona's *St. Paul Recovering his Sight* in the Capuchin Church, but he was never entirely at ease with the austere High Renaissance and Baroque paintings he saw in Rome. Natoire was perceptive enough to see where his pupil's real bent lay. 'Fragonard, with his gifts, has an astounding facility to change his ways from one moment to the next, and this makes his work uneven' (30 August 1758). The one word which crops up most often in these letters in reference to Fragonard is 'feu', an ardent, spontaneous temperament which his mentors wished to encourage but, with their routine habits and half-hearted praise of his work, did their best to quench. We are reminded of Stendhal's comment in chapter LXVI of his *Histoire de la peinture en Italie* that 'Les règles boiteuses ne peuvent

fig. 2 Richard de SAINT-NON
Illustration from 'Fragments des peintures et tableaux les plus intéressants des palais et églises d'Italie' after Saint-Non's own pen and ink sketch from 'Peinture antique d'Herculanum'
Etching 13·8 × 19·8 cm.

fig. 3 Hubert ROBERT
Villa Conti at Frascati
Oil on canvas 62 × 47 cm. Besançon, Musée des Beaux-Arts

Probably executed in or around 1760 during Robert's stay with Fragonard and Saint-Non in Rome and the neighbouring countryside. This painting shows how close Robert and Fragonard were, except that Robert's style was rather more detailed and meticulous.

suivre l'élan du génie'. 'Lame rules cannot keep up with the impetus of genius.'

In November 1759 Fragonard was fortunately rescued from this state of disorientation by the timely appearance in Rome of the abbé de Saint-Non. Richard de Saint-Non (1727–1791) was typical of the eighteenth-century amateur artist and patron. As abbé commendataire of Pothières, near Langres, he possessed a handsome ecclesiastical sinecure worth 8000 francs a year income, which allowed him to devote himself to the arts. He was a reasonably skilled draftsman, a good etcher (fig. 2), later etching many of Fragonard's own drawings, and travelled widely in England and Holland, making engravings after Rembrandt. Saint-Non deserves to be remembered above all for his generous support of young artists, especially Fragonard, for whom he performed the same service which the Abbé Gougenot had for Greuze, enabling him to travel and complete his artistic education. His assistance was especially good for Fragonard because he had none of that academic pedantry and exaggerated respect for classical Antiquity of Caylus and Winckelmann. Like the président de Brosses he belonged more to the generation of the 'picturesque' traveller, interested in Italy for its way of life more than its monuments, an observer with a sharp eye for detail and the happy accident. The abbé arrived in Italy with another protégé, the young painter Taraval, and presumably met Fragonard early in 1760. It was probably through Saint-Non that, about this time, Fragonard met Hubert Robert. The two artists became inseparable friends, and their paintings of the Italian countryside are sometimes so close as to be indistinguishable from each other (fig. 3). Hubert Robert, of a comfortably off middle-class family, had all the advantages which Fragonard lacked, a sound classical education and knowledge of literature. It was very likely that in contact with Robert and Saint-Non, Fragonard became familiar with the great Italian poets, Tasso and Ariosto, whose works he so brilliantly illustrated (Plates 39a–f). We can imagine the three friends reading passages from Ariosto's *Orlando Furioso* together, and although Fragonard was not bookish he showed a rapid, intuitive understanding of the poem's meaning.

In March 1760 Natoire gave Fragonard leave of absence to travel with Saint-Non to Venice, making several stops en route. Their first port of call was Tivoli, where the Duke of Modena put the Villa d'Este at their disposal. There they spent the summer of 1760, joined by Robert, talking and sketching from morning to dusk in the decaying splendour of the park with its avenues of cypresses, its fountains and statues overshadowed by the massive façade of the villa. This was a decisive moment in Fragonard's career, for at last he could forget the vexations of academic teaching and give free rein to his own inclinations. The result was that rapid, sketchy and yet comprehensive vision of nature in *The Giant Cypresses at Tivoli*, *The Entrance to Fontanone* and the *View of the Coast near Genoa* (Plates 8, 9 and 10). It was at Tivoli that he learned to draw the huge trees and their overhanging boughs which dominate the later Parisian park scenes. If these seem exaggerated to modern observers it should be remembered that in the eighteenth century the gardens of the Villa d'Este, and most other Italian gardens, were in a state of considerable neglect; overgrown trees and toppling statues were not simply the product of a lyrical imagination. The président de Brosses, writing in 1737–40, reported that the whole place was already decayed. 'The Duke of Modena neglects it totally; the gardens, arbours, woods, the sloping and terraced flowerbeds have all gone to seed and lie fallow' (*Lettres familières écrites d'Italie*).

After this idyllic sojourn at Tivoli the party presumably went on to Venice. Although there is no actual record of a visit, it seems inconceivable that Fragonard had not seen at first hand frescoes by Giovanni Battista Tiepolo, his closest counterpart among contemporary painters. In March 1761 he set off alone, at Saint-Non's expense, for Naples, where he was most deeply impressed by the works of Solimena and Luca Giordano, both *fa presto* artists with whom he felt a closer affinity than with the High Renaissance painters. Meanwhile Natoire seems to have been reconciled to his pupil's new-found freedom and wrote to Marigny promising that 'The Abbé [Saint-Non] will bring back with him a number of the young artist's charming sketches'. This refers, we may suppose, to Fragonard's drawings of Tivoli. During this period Fragonard had acquired a much greater ease with the Old Masters, as we can see from the wonderful pen and ink sketches after works in Bologna, Rome and Naples incorporated into Saint-Non's illustrated volume, *Fragments des peintures et tableaux les plus intéressants des palais et églises d'Italie* (1770–73). This work consists of ninety pieces, engraved aquatints by Saint-Non from drawings by Robert, Fragonard and Saint-Non himself. The abbé's other large volume, the *Voyage de Naples et de la Sicile* (1781–86), also included etchings after Fragonard's drawings, but much of the descriptive text is not by Saint-Non, who never travelled as far as Sicily, but by Vivant Denon. This vast and costly enterprise, originally designed to cover the whole of Italy, had to appear in truncated form and was to be the ruin of Saint-Non, whose financial position was not improved by the suppression of his abbey by the Revolutionary government in 1791. As an admirer of Rousseau and Voltaire, Saint-Non, however, shared the Encyclopédistes's radical sympathies. He quickly reconciled himself to the new situation and does not seem to have shed many tears over the passing of the Ancien Régime. The best testimony to his intuitive understanding of Fragonard are the words he wrote to his brother, M. de la Bretèche. 'M. Fragonard is all ardour; his drawings are so numerous that one cannot wait for the next; they delight me. I find a kind of magic in them.'

Fragonard returned to Paris with Saint-Non in September 1761. The next four years of his life are sparsely documented, but he must have spent a good deal of this time preparing for his début as a history painter with the exhibition of *Coresus Sacrificing Himself* in 1765. It must also have been shortly after Fragonard's return from Rome, between 1762 and 1765, with the vision of Tivoli fresh in his memory and armed with numerous sketches, that he painted the group of Italianate landscapes including *The Washerwomen* (Plate I), *The Gardens of the Villa d'Este* (Plate 12) and *The Waterfall at Tivoli* (Plate 11). All of these are minor masterpieces, quiet, restrained, with a remarkable density of texture, and rigorous in composition. The use of thick chalky white paint in *The Washerwomen* serves to emphasize the stiffness of linen hanging out to dry. Fragonard's feeling for surface texture was as unerring as Chardin's. The symmetry of these paintings, for example *The Gardens of the Villa d'Este* with its carefully balanced statues and terraces, was given by the subject, but they make an interesting contrast to such non-Italianate landscapes as the slightly earlier *Storm* (Plate 6), of about 1759, and the *Annette and Lubin* (Plate 7), dating from soon after 1762. In both of these the mass of the composition is heaped up to the left of centre. In *The Storm* this creates a tumultuous, dramatic effect heightened by the dark clouds on the right, while in the pastoral *Annette* the brown hillock contrasts with the airy blue spaciousness of the distance. Fragonard was not merely a factual observer of nature; he also portrayed the full range of its moods. In the group of 'Dutch' landscapes, for instance, the *Autumnal Landscape* at Amiens, the scene at Grasse (Plate 20) and, the most poignant of all, *The Three Trees* (Plate 19), this can be as desolate and melancholy as in any by Ruysdael. Here nothing could be more remote from the Provençal *joie de vivre* which Fragonard is supposed to typify. This raises the problems of Fragonard's possible visit to Holland. If he went there at all, and there is no evidence that he did, it may have been during the years 1761–65, although M. Ananoff suggests that it was a decade later, in 1772–73. The hypothesis of a Dutch visit rests solely on visual affinities with Rembrandt, but these could be explained from Fragonard's knowledge of works in French collections. Saint-Non had also made a number of etchings after Rembrandt which Fragonard would have known.

Official recognition and fame came for Fragonard in March 1765, when he submitted *Coresus Sacrificing Himself* (Plate 17) as his 'morceau de réception' to the Academy and was received as a full member of that body, the highest honour to which an eighteenth-century artist could aspire. What is remarkable is that when this highly dramatic work was exhibited at the Salon of the same year it managed to satisfy both the jury and the critics, whose views were usually diametrically opposed. Diderot was so excited by the painting that he allowed his fertile imagination to re-create it in the form of a dream ('Plato's Cave'), in which he multiplied the number of victims and laid so much stress on the macabre element ('l'effroi') of the work that he nearly transformed it into a Gothic horror novel. Diderot was perhaps guilty of only slight exaggeration, for there is something unmistakably nineteenth century in the Furies swooping down on the left, almost anticipating Prudhon's *Justice and Vengeance Pursuing Crime*, or in the figure of the young priest to the left of Callirhoe, with his tablet on his knees, who has the same expression of inspired zeal as Ariosto in the later drawing (Plate 39a). Whatever latent Romantic tendencies the art historian may discover in Fragonard, they were, of course, unconscious on the painter's part. In the context of its own time, Fragonard's work marked a notable and successful attempt to reconcile the two conflicting tendencies which divided French eighteenth-century art, the Poussiniste demand for a competent and dramatic treatment of the set theme and the Rubéniste love of colour and variety for its own sake.

Fragonard, however, seems to have been dissatisfied with his own success. Although Cochin persuaded Marigny to have the *Coresus* reproduced by the Gobelins tapestry works, payments from the royal treasury were notoriously slow and ungenerous. The inadequate financial reward accorded to the honour of history painting, coupled with the feeling that this kind of work went against his natural inclinations, persuaded Fragonard to turn to a different clientèle, the rich financiers, aristocratic amateurs and courtesans with bottomless pockets and easily satisfied tastes. They were not the sort of people to take much notice of the new fashion among intellectuals for Antiquity and morally elevating subjects. From now on M. de Varanchan, the marquis de Véri, Randon de Boisset, Sophie Guimard and, later, Bergeret de Grancourt, provided Fragonard with a steady stream of commissions for erotic nudes, cupids, unmade beds, lively haystack scenes and similar subjects, all dashed off in a few rapid strokes of the brush. They gave him a quick source of income at the cost of no great mental exertion. Some of these, like *The Longed-for Moment* (Plate 30), *The Stolen Shirt* (Plate 31), or, most suggestive of all, *The Bed with Cupids* (Plate 33) are deservedly famous specimens of erotic art, what the Goncourts called 'the divine poem of Desire'. They are also strangely de-personalized creations. Fragonard's lovers have no individual identity, and their faces consist only of slot-like eyes and pink cheeks. Sexuality for him was simply violent movement, a manifestation of the universal 'élan vital', without any of the psychological undertones of Watteau and equally differ-

ent from the langorous passivity of Boucher's love scenes. It could sometimes stop at mere titillation, as in *The Swing* (Plate II), painted for the baron de Saint-Julien, who is seen lolling in the foliage on the left enjoying a good view of his young mistress's legs. To an unsympathetic observer this painting might easily be held up as an example of the frivolous, licentious way of life of the upper classes in the Ancien Régime. Love is reduced to idle gallantry, and marriage is derided, for the husband is giving a helping hand to the swing. The girl is a doll-like coquette dressed up in flounces and billows, while the statue of Cupid archly puts his finger to his lips; and yet there is a peculiar intensity in the expression of the baron's face which reminds us that the age of 'sensibilité' had already dawned. This little episode is somehow engulfed by a luxuriant hot-house growth of vegetation, reminiscent of Zola's Le Paradou in *La Faute de l'Abbé Mouret*, another erotic fantasy world created by the Provençal imagination.

Meanwhile C. N. Cochin, the artist and critic, continued to exert his influence in Fragonard's favour and in 1766 got for him the important commission to paint the allegorical figures of *Day* and *Night* for the royal Château de Bellevue. Cochin by that time was firmly established as artistic mentor to Marigny, who, as Directeur des Bâtiments from 1754, was responsible for royal patronage, a position of considerable power. In 1749 Cochin had accompanied the young Marigny, when still M. de Vandières, on an educational tour of Italy during which he succeeded in weaning his pupil from a preference for the 'petite manière', the Rococo mode of Boucher and Van Loo, in favour of the grand style. Cochin was no doctrinaire, however. He was among the earliest French critics to state the heretical view that Raphael was perhaps not the greatest painter ever to have lived and was also a pioneering champion of Veronese and the colourist Venetian school. In 1749 he wrote prophetically that 'The true charm of nature which consists in colour, harmony and the overall concordance of the picture has yet to be discovered'. Elsewhere, when he wrote of 'nature in its infinite variety', he seems to make a direct appeal for an artist like Fragonard to translate his wishes into reality. In December 1766 he wrote to Marigny suggesting a competition designed to discredit Boucher, Pierre and Vien and to throw his favourites, notably Fragonard, into prominence. There is no record of the outcome of this scheme.

In the following year, 1767, Fragonard exhibited several works at the Salon, among them *Groups of Putti in the Sky*, the *Head of an Old Man* and a number of drawings. This time, however, he failed to win the critics's approval. Diderot, who devoted his longest and most inspired commentary to this Salon, scornfully dismissed the *Putti* as a 'fine, soft, yellow, well-browned omelette'. He clearly suspected Fragonard of deserting the grand manner which he had seemed to inaugurate with the *Coresus* in 1765 and of returning to Rococo mannerism: 'M. Fragonard, when a man has made a name for himself, he should have a little more self-respect'. Bachaumont similarly accused him of back-sliding into the 'luxuries of Capua'. In these and other criticisms there is the strong implication that private patronage had been a bad moral influence on Fragonard, and that instead of painting 'bambochades' for rich financiers, he ought to have been devoting his energies to the revival of history painting, which the state was doing its best to promote. The critics's disappointment was confirmed by Fragonard's failure to exhibit at the following Salon of 1769, when the newspapers began to report that 'the lure of profit and the interest in boudoirs and wardrobes have diverted the painter from striving after glory'.

By this time Fragonard had reached his maturity. It was in or around 1769 that he painted some of his most idiosyncratic works, the celebrated 'fancy portraits' (Plates III and 25–29) which must rank among the most extraordinary virtuoso pieces in French art. Two of them, we learn from Fragonard's own inscriptions, were painted in only one hour; more than anything else they have created his reputation as a *fa presto* artist. The painted surface is reduced to long shreds of colour which seem to perform arbitrary arabesques at the caprice of the artist. Whereas most French portraiture is usually sober and restrained, these are unashamedly extravagant, almost Spanish in their self-confident assertiveness. They seem to thrust themselves forward at the spectator, like *The Warrior* (Plate 27) with his reckless contempt, a proud isolated individual framed only by a stone balcony and set against a bare back-ground. This is an essentially aristocratic art (two of the sitters are members of the Harcourt family) in keeping with the mood of one of Fragonard's favourite books, *Don Quixote*. It is no accident that the abbé de Saint-Non is shown in the guise of Cervantes's hero (Plate 29). While some of the sitters can be identified, for example Diderot, Sophie Guimard and Saint-Non, not all are clear. Who, for instance, is the mysterious *Warrior*, with his proud, angular jaw, embittered expression and clenched fist, seeming to defy the world like some tragi-comic hero?

On 17 June 1769, after thirty-seven years of the care-free bachelor's life, Fragonard married his pupil Marie-Anne Gérard, the elder and less attractive of two sisters from Provence. The younger sister, Marguerite Gérard (Plate 45) came to live with the couple in 1775 and, although barely literate, she was a highly intelligent woman and a talented artist in her own right (fig. 4), collaborating in many of Fragonard's late works. An extensive correspondence between her and Fragonard survives, quoted extensively by Portalis, which suggests that, if she did not actually become his mistress, the

fig. 4 Marguerite GÉRARD
The Reader
Oil on canvas 65 × 54 cm. Cambridge, Fitzwilliam Museum

The standing woman is wearing a cream satin dress, the seated woman a red jacket with white fur. The background is buff grey, with dark green curtains and table cloth. The figure of the standing woman bears a close resemblance to the girl in Fragonard's *Stolen Kiss* in the Hermitage, which may also be partly by the hand of his former pupil, Mlle. Gérard. *The Reader* was bequeathed to the Fitzwilliam Museum by Charles Brinsley Marlay in 1912.

painter may later have regretted his choice of wife. Biographers have made a good deal of the influence of marriage on Fragonard's art. It is true that after this he painted a considerable number of scenes of happy families and children at school like *The Cradle* (Plate 50), *The Schoolmistress* (Plate 51) in the Wallace Collection and many others, which certainly reflect his own domestic contentment and delight in children. Fragonard was, however, a canny enough opportunist to realize that by 1770 the moral tide had already turned against licentious subjects, even though he continued to turn them out when the occasion demanded. Diderot had acclaimed Greuze as his ideal painter of the simple moral virtues; the whole of the literate French public was reading the novels of Richardson and, in the theatre, the serious-minded *drame bourgeois* had ousted the pastorals of Favart and Lamothe. It may well be, then, that Fragonard's new interest in domestic subjects was an attempt to rival Greuze on his own ground and to exploit an increasingly popular vein. It is very likely that the subject of *The Cradle*, and other paintings based on the theme of the happy family, is taken from a now forgotten novel *Miss Sara Th . . .* (1765) by J. F. de Saint-Lambert, a tale of a rich, well-born young woman who voluntarily abdicates her social position by marrying a farmer for love. They lead a model life of conjugal felicity, beget five children and maintain a simple, orderly household in which the servants are treated as equals.

Meanwhile, despite the disapprobation of the critics and the change in the moral climate, Fragonard continued to be the most sought-after artist for the decoration of Parisian drawings rooms. It was probably in 1770 that he was commissioned to paint the panels for Mlle. Guimard's house, built by Ledoux, on the Chaussée d'Antin. Sophie Guimard (1743–1816) was the most famous ballet dancer of her time. Bachaumont described her as being 'as light as Terpsichore' (May 1762), and her slender, willowy form is admirably caught in Fragonard's portrait (Plate 28). When Fragonard knew her around 1770 she was kept in the height of luxury by the prince de Soubise. Everything she did and touched set the fashion in Parisian society and filled the gossip columns of the newspapers. She gave three dinner parties a week, one for grands seigneurs and members of the court, another for artists and writers; the third was simply an orgy where, in Bachaumont's words, 'debauchery and vice were carried to the extreme'. Her reputation for elegance and luxury spiced with scandal, which earned her Marmontel's epithet 'la belle damnée', made Fragonard the obvious choice of painter for the decoration of her hôtel. She installed her own private theatre, with a ceiling by Taraval, in the house, opened on 8 December 1772. Fragonard must presumably have worked for her between 1770 and 1773. In 1772 Grimm reported that the work was nearly finished and commented that 'Love bore the expense and pleasure designed its plan; no finer temple was ever erected in Greece in honour of this goddess. The drawing room consists entirely of paintings. Mlle. Guimard is depicted as Terpsichore . . . .' Then, in 1773, for some unknown reason, Fragonard fell out with Mlle. Guimard. There were inevitable rumours about his relationship with the dancer, and the motive may have been one of personal spite but, more likely, he had grown tired of her haughty, capricious behaviour. The Goncourts report, on Grimm's authority, that to revenge himself Fragonard changed the portrait of her from Terpsichore to the Fury Tisiphone. Although several portraits of Mlle. Guimard by Fragonard survive, there is no trace of the one of her as Terpsichore, mentioned by Portalis as being in a private collection. In any case, Fragonard left the work unfinished and handed it over to the young

Jacques-Louis David. In 1786 Sophie Guimard, heavily in debt, was forced to put her house up for sale by lottery and, after passing through various hands, it was demolished in the nineteenth century.

Fragonard's second and much more important commission in 1770 was to paint a series of decorative panels and overdoors for Mme. du Barry's new pavilion, also by Ledoux, at Louveciennes. These are the panels known as *The Pursuit of Love* (Plates 34a–e), now in the Frick Collection, New York. The story behind this commission and the reason for Mme. du Barry's sudden refusal of Fragonard's panels in 1773 are complex and still not quite clear. Mme. du Barry (1741–93) succeeded in 1769 to the position of Louis XV's recognized mistress, left vacant since the death of Mme. de Pompadour. Du Barry also assumed Pompadour's unofficial role as patroness of the arts, but she had none of her predecessor's artistic and intellectual gifts. Mme. de Pompadour could sketch, draw and play the harpsichord, and when Boucher painted her portrait he showed her surrounded by all the attributes of the arts. Mme. du Barry, who was no less prodigal of the treasury's funds, spent more money on jewellery and clothes than on works of art. She was, however, very anxious to keep abreast with the latest artistic fashions, and this may well be the reason why, having discovered that Fragonard's Rococo style had passed from favour in certain advanced circles, she returned his canvases and handed on the commission to Vien, whose work was also more in keeping with the Neo-Classical exterior of the building. A diluted Etruscan form of Neo-Classicism, suited to the small scale of Parisian drawing rooms, had become the rage in fashionable society and Grimm reports that 'tout est à la Grecque'. 'Everything is in the Greek style.' Fragonard, with his 'shepherds's romances', as Bachaumont slightingly referred to *The Pursuit of Love*, suddenly found himself outdated overnight. Today, however, the pendulum of taste has swung back again, and it needs some effort of the historical imagination to see why Mme. du Barry should have preferred Vien's pleasant but anemic scenes to Fragonard's overwhelmingly beautiful panels.

As their title implies, the subject of the series is the triumph of love, unless the fifth panel, *Abandonment*, also belongs to the sequence. This seems unlikely as Fragonard could hardly have wished to inflict an unhappy ending on his patron. Bachaumont suggested that *The Pursuit of Love* was an allegory on the mistress of the house, and they may allude to the love between Mme. du Barry and Louis XV; some critics have noted that the lover has the king's features. Both the lovers are dressed in the height of fashion, and the scenes are set in an opulent, overgrown garden, with a profusion of urns, columns and shrubs, a little walled enclave threatened by a riot of natural growth. Fragonard treated the scenes like four episodes from a contemporary novelette, but, as in *The Fête at Saint-Cloud*, by the sheer exuberance of his imagination he transformed what might have been vignettes on the scale of Gravelot or Eisen into a work of epic poetry. The lightness and exquisite poise of the figures, and the fact that each episode has a different *mise en scène*, suggest a possible source of inspiration in contemporary ballet. The third panel, *The Declaration of Love*, has all the elegant artificiality of a reconciliation scene between lovers. The girl holding the wreath in the fourth scene has very similar features to the portrait *Study* in the Louvre which bears some resemblance to Sophie Guimard. The sequence of the pictures is easy to read and their meaning plain. In the first, *Storming the Citadel* (Plate 34a), an eager lover clambers up a ladder and over the low wall, disturbing the girl, who seems genuinely startled by the intrusion. In *The Pursuit* (Plate 34b) he comes bearing a rose and breaks up a party 'à deux'; this time the girl trips away coquettishly like a ballet-dancer, but is not displeased. By the third, *The Declaration of Love* (Plate 34c) the lover has already won over his sweetheart, and they exchange a tender message while the statue seems to give her maternal blessing to their union. Finally, in *The Lover Crowned* (Plate 34d), the young man is crowned by the girl with a wreath, and her musical instruments lie scattered on the ground as if to symbolize the victory of love over the arts; meanwhile a young artist is busy sketching the scene for posterity. The feelings and décor of *The Pursuit of Love* are those of Beaumarchais or the young Mozart, for they epitomize the all-absorbing subject which love had become to the late eighteenth century, not 'l'amour passion', with its attendant suffering, but 'la galanterie', the pursuit, the chase and the conquest.

At roughly this point another important patron stepped into Fragonard's life, Bergeret de Grancourt (1715–85). Bergeret was Treasurer-General of Finance for the region of Montauban and the son of a member of that unpopular class of men, the Fermiers-Généraux, the tax collectors under the Ancien Régime. Bergeret was also an immensely rich man, the owner of the château and estate of Nègrepelisse in Languedoc and other properties near Paris. To judge by Vincent's portrait of him (fig. 5) painted in Rome in 1774, Bergeret wearing a turban and in silk pyjamas, he was the epitome of the bloated financier, arrogant, ruthless and opinionated. He had none of Saint-Non's finesse and generosity, but his interest in the arts was genuine. Such a man's luggage was not complete without a painter, and in October 1773 he invited Fragonard and his wife to accompany him to Italy. Bergeret wrote an account of their journey in a series of letters to his nieces in Paris. The party travelled via Uzerches, Brive, Nîmes and Aix-en-Provence, where they admired some fine paintings in the collection of the président d'Albertas, embarked on a boat at Antibes for Genoa, then

fig. 5 François-André VINCENT
Portrait of Bergeret
Oil on canvas 64 × 37 cm. Signed: Vincent f. Rome 1774. Besançon, Musée des Beaux-Arts

François-André Vincent (1746–1816) later became a prominent history painter. At the time he painted this portrait of Bergeret he was a pupil at the Ecole de Rome.

went on to Pisa and Florence. Prompted no doubt by Fragonard, Bergeret constantly reminded his readers of his visual preoccupations, 'Nous voyons tout en tableau', and he made careful notes of his impressions of people, scenery and works of art. At the beginning he was delighted with Fragonard's company, found him an easy travelling companion and praised him as a 'painter of excellent talent, whom I need particularly in Italy', but after this there is very little mention of Fragonard, perhaps suggesting that Bergeret failed to appreciate the artist's true value. The climax to their journey was Rome, where they arrived on 5 December. Bergeret was so overwhelmed by the city that his journal breaks into a series of exclamations at the sight of so many palaces, fountains, ruins and marbles. There are some interesting pages on the Ecole de Rome, and Bergeret could not find enough praise for its pupils, Vincent, Berthélémy, Suvée, Taillasson and Ménageot. The party was taken on guided tours of Rome by the architect P. A. Pâris, who later inherited many of Fragonard's best drawings and gave them to his native Besançon.

fig. 6 Gabriel de SAINT-AUBIN
The Fête at Saint-Cloud
Black chalk, brown and grey wash 21·3 × 29·9
cm. Edinburgh, National Gallery of Scotland,
Department of Prints and Drawings

In the lower right-hand margin and on the
back the drawing is inscribed: Pater. This is
almost certainly not authentic.

Then, early in 1774, they visited Naples and from there set back on the long journey home via Venice, Vienna, where they saw the Prince of Liechtenstein's gallery, Dresden, Strasbourg and, finally, Paris. On their return the strain of travelling together for a long period broke out in a violent quarrel between Fragonard and his patron. Bergeret struck out his earlier words in praise of the artist and demanded the right to keep all the drawings Fragonard had made during the journey as payment in lieu of expenses. The case was then taken to court, where it was decided that Bergeret should return Fragonard's drawings or pay him substantial compensation. Whatever the rights and wrongs of this incident it clearly reveals Fragonard's independent attitude towards his patrons, his confidence in his status as an artist and his refusal to be treated as a lackey.

It was probably soon after his return from Italy, in 1775, but possibly earlier, that Fragonard was asked by the duc de Penthièvre to decorate his dining room in the Hôtel de Toulouse, now the Banque de France. The result was *The Fête at Saint-Cloud* (Plates 42 and VII), Fragonard's masterpiece and one of the least known and most beautiful of all French paintings, the late eighteenth-century pendant to Watteau's *Embarkation for Cythera*. The setting is the park of the Château de Saint-Cloud, just outside Paris, which then belonged to the royal family but was freely open to the public. This is what the countryside meant to the metropolitan artist and his public at the time, just as Montmartre was country to Nerval and Asnières to Renoir. About 1750 the French attitude towards the country began to change. After a century of metropolitan life centred on the court people started to hanker after the rural retreat. 'Tout le monde est maintenant à la campagne.' 'Every-

body is in the country today' wrote the Abbé Raynal in October 1750. Nobles and financiers spent up to a previously unthinkable six months on their estates, bourgeois bought up villas in the Ile de France, poets retired to their 'ermitages', while the working population had to be content with a day out in the country. Saint-Cloud might be compared to Vauxhall Gardens in eighteenth-century London and provided the same kind of diversions for the people of Paris. Every year, on the last three Sundays of September, a festival was held when the fountains played and entertainment was provided by actors, dancers, fireworks, booths and puppet shows. This custom was maintained after the Revolution, when the park became state property. Not only the fashionable visited Saint-Cloud; it was also the favourite rendez-vous of artists like Gabriel de Saint-Aubin (fig. 6) and Hubert Robert, who found a ready-made subject in its varied spectacle. By far the most numerous visitors were the petits bourgeois and artisans of Paris, the shop-girls and dressmakers who flocked there in boatloads down the Seine and then crowded on to carts for the rest of the journey. After a day's riotous entertainment and a supper of vinegary wine and badly cooked meat at exorbitant prices, they would make their way happily home through the woods. These are the spectators, graphically described in L. S. Mercier's *Tableau de Paris*, we see in Fragonard's picture. Some are grouped round the puppet show, some are playing a game something like roulette, others are watching the actors, while a few are content to loll idly with their sweethearts (Plate VII). The whole picture is composed of a series of delightful vignettes and yet is conceived on a vast, epic scale, unified by the enormous Italianate trees which seem to dwarf the human figures and reduce their activities to mere antics. All natural growth is larger than life; even the fountain is denser and more powerful than a real one. Does this mean that Fragonard has invested nature with superhuman powers and produced a kind of pantheistic vision close in spirit to Rousseau? He has been considered as a precursor of the pre-Romantic sensibility in which ominous, uncontrollable forces threaten to reduce mankind to insignificance; witness the broken capital in the centre and the menacing presence of the statue on the right, like the Commendatore in Mozart's *Don Giovanni*. Are these symbolic of decadence, or are they simply the compositional devices which Fragonard, like Hubert Robert, found irresistible?

This streak of Romanticism seems, at first sight, even more predominant in *The Fête at Rambouillet* (Plate VIII), with its gnarled Salvatorial tree in the centre and its foaming, rocky waters like an eighteenth-century version of Poussin's *Deluge*. There is, however, nothing really sinister in this picture; it is simply a party enjoying the thrill of pretending itself out in a storm at sea, an elaborate charade designed to give the public the same kind of vicarious frisson which scenes of ship-wrecks by Vernet and Loutherbourg gave to city-dwellers who had never been near the sea in their lives. This does not, of course, argue against Fragonard's Romanticism, indeed this kind of flirtation with danger and quest for substitute experience is one of its most deeply-rooted characteristics.

With these two great decorative park scenes Fragonard at last came into his own. They are his unique personal creation. He had evolved a genre which, if not entirely new, was to stand him in good stead for the next decade and made a major contribution to French art. It was also in the 1770s that he painted the marvellous panels in Washington, *The Game of Hot Cockles, Blind Man's Buff* (Plate 35) and *The Swing* (Plate 36). Building upon his empirical observations of nature in Italy, Fragonard magnified his chalk sketches and translated them into a new modern idiom which reflected contemporary French life with the same delicate precision and imagination as Watteau. The Goncourts were right to see Watteau and Fragonard as the two poets of the eighteenth century, for there is always an element of the fantastic, for example the weird shape of the mountain in the distance of *The Swing*, which lifts Fragonard far above the minute, literal transcription of outdoor scenes by Gabriel de Saint-Aubin (fig. 6), who, although he may, in a narrow sense, have been the inventor of the genre, nevertheless, remained a superficial observer. While all the details in *The Swing* are perfect in themselves as pieces of realistic observation, the woman playing with the dog, another woman peering through a telescope into the distance, they are only the elements in Fragonard's epic vision of landscape. For the first time, perhaps, in the eighteenth century he raised decorative painting to the level of great art.

In all Fragonard's best work there is this fine balance between fantasy and reality. It was, understandably, the naturalistic side of him which attracted the Impressionists, especially Renoir, who looked to Fragonard as a pioneer in the art of showing people reading, musing, or doing nothing in particular. They felt a special affection for paintings such as *The Music Lesson* (Plate 22) and the *Woman Reading* (Plate 46), which seem to distil the very essence of French domestic life in the eighteenth century. This affinity, which extends to colour and texture, is particularly close in Renoir's female portraits, while his *Grandes baigneuses* of 1887 can be seen as a distant reply to Fragonard's own *Bathers* (Plate 18). For the late nineteenth century Fragonard seemed like a precursor of their notion of the 'painter of modern life', a recorder of the way of life and leisure pursuits of a particular society. There is an interesting reflection of this view of Fragonard in Marcel Proust's *La Prisonnière*, where the narrator tries to telephone Andrée but only succeeds in arousing the fury of 'implacable deities'. 'As I waited for her to finish her conversation, I asked myself how it was—now that so many of our painters are seeking to revive the feminine portraits of the eighteenth century, in which the cleverly devised setting is a pretext for portraying expressions of expectation, spleen, interest, distraction—how it was that none of our modern Bouchers or Fragonards had yet painted, instead of "The Letter" or "The Harpsichord", this scene which might be entitled "At the Telephone", in which there would come spontaneously to the lips of the listener a smile all the more genuine in that it is conscious of being unobserved.' The nineteenth-century nostalgia for the eighteenth century has rarely been better expressed, for Proust, like the Goncourt brothers, believed that the painters of the Ancien Régime were a greater credit to French art than all the 'doctrinaire' artists of the Revolution put together.

Fragonard was, finally, one of the most brilliant literary illustrators of all time. In his drawings for the *Orlando Furioso* (Plates 39a – f) and *Don Quixote* (Plates 48a, b) he revealed himself equal in creative power to the original writers, and his rapid understanding sent him straight to the meaning of a text. Examples of French literature from his own time do not seem to have fired Fragonard's imagination to the same degree. He illustrated subjects from Marmontel, from Mme. de Genlis's *Les Veillées du château* and from La Fontaine's *Contes* (Plates 57a, b), but they are hardly his most inspired creations. He was no intellectual, and his contacts with the major writers and thinkers of his own day were minimal; there is nothing to show that writers thought very highly of him in return. And yet Fragonard's art seems inseparable from the eighteenth-century mould of thought. His great decorative scenes, after all, coincide with the celebration of natural creation in Rousseau and Bernardin de Saint-Pierre, while many of his 'galant' subjects provide an almost exact complement to the poems in Parny's *Poésies érotiques*. His closest affinity is perhaps with Diderot, and it is more than a coincidence that the great collector Hippolyte Walferdin associated the writer and the artist in a twin cult. Fragonard's art has all the contradictions, the variety, exuberance and flexibility of Diderot's prose, the same alternating flights of fancy and earthy Rabelaisian humour. He had too the same unreflecting, spontaneous creative urge as Diderot, the same Protean mobility. Such literary parallels should not be pushed too far, however, even in an age when literature and painting tended to go hand in hand. For where Fragonard was most original and least like other eighteenth-century painters was in his awareness of the artist as a unique being, subject to that irrational element, inspiration. Anyone wishing to show that 'genius' was the invention of the late eighteenth century might well find proof in drawings like *Ariosto Inspired by Love and*

Folly (Plate 39a), the *Homage to Gluck*, *The Poet's Inspiration* or *Plutarch's Dream* (Plate 16). These strange drawings are all clearly related in theme and style and perhaps tell us more about Fragonard's ideas, tastes and view of himself than any other work. They form a kind of miniature Pantheon; each one is dedicated to an artist-hero, who sits contemplating some fantastic vision in his mind's eye. The drawing of Ariosto, which might possibly have served as a frontispiece to Fragonard's projected edition of the *Orlando Furioso*, shows the poet crowned with laurels sitting barefoot on a stone bench gazing at the two presiding geniuses of his poem, Love and Folly. The next, *Homage to Gluck*, with a poet seated in front of three busts of Gluck, Homer and Virgil and a slab inscribed 'AMANTI DEGLI (sic) ARTI', was probably intended to celebrate the performance in Paris of *Iphigénie en Tauride* in 1779. It was followed by *The Poet's Inspiration*, or *Homage to La Fontaine*, showing the same poet in front of busts of Erato, La Fontaine and Rousseau, and finally by *Plutarch's Dream* (Plate 16), a figure seen reading a copy of Plutarch's *Lives of Famous Men*, with scales, one balance bearing the inscription 'Grandeur', the other 'Médiocrité'. These four apparently insignificant drawings reveal Fragonard as a man of strong intellectual sympathies, with a quasi-Romantic belief in the poet, musician or painter as a visionary, blessed with a divine gift. It only remained for writers to state this idea in words for it to become common currency in the nineteenth century.

The remainder of Fragonard's career can be briefly told, for although he lived on until the age of seventy-four, his period of greatest creativity was over. In 1780 his son, the painter Alexandre-Evariste, was born at Grasse, and in 1788 he was severely shaken by the death of his only daughter Rosalie. Meanwhile the newspapers published a steady stream of engravings after his most popular compositions, *The Bolt* in 1784, *The Fountain of Love* in 1785 and *The Stolen Shirt* in 1787, all of which attests to Fragonard's continuing popularity with the general public. It was also after 1780 that he began to paint the allegorical, mythological works which have earned him the epithet 'a Romantic before the Revolution'. One of these is entitled *Homage Rendered to Nature by the Elements* (now lost, Wildenstein cat. no. 490), or, more poetically, *Le réveil de la nature*, and shows a draped and hooded statue of Nature on a pedestal, surrounded by a flurry of cupids, eagles, doves and a pair of fawning lions at her feet, the whole of nature come to pay homage to its source. On paper this sounds a typically eighteenth-century subject, but there is something artificial and fantastic about it which looks forward to Prudhon or Girodet. There is also an unmistakably nineteenth-century quality about *The Fountain of Love* (Plate 58), with its two lovers drawn as if by magnetic force to the fountain, the symbol of desire. Although they are supposed to convey violent move-

ment, the figures are linked in a symmetrical, statuesque outline more suggestive of Neo-Classical immobility. Even more like Prudhon is the strange, allegorical *Sacrifice of the Rose* (Wildenstein cat. no. 497) showing a fainting, naked woman, one of love's victims, with a cupid burning a symbolic rose on the altar of sacrifice. These curious works, which may have obscure literary origins, indicate that Fragonard's mind had already taken on a distinctly Romantic cast well before such subjects became popular in the nineteenth century. This does not mean that he renounced his 'galanteries', as we can see from the late *Stolen Kiss* (Plate 55) or *The Bolt* (Plate 56); only that his style, with the direct or indirect help of Marguerite Gérard, had undergone a subtle modification. The chubby, naked bodies of the earlier erotic pieces gave way to the tall willowy form of the girl in *The Stolen Kiss*, who snatches at her satin scarf and lets herself be furtively kissed while her parents or relatives next door carry on with their card game unawares. The whole scene is one of the most perfect late eighteenth-century interiors ever painted.

In 1789 Fragonard was overtaken by the Revolution. As the protégé of not particularly enlightened members of Ancien Régime society he can hardly have welcomed the event, but with his customary flexibility he managed to survive the worst years and even modestly to prosper. In an effort to ingratiate themselves with the new authorities, his wife and Mlle. Gérard offered their jewels to the National Assembly in 1789. Then in 1790 Fragonard prudently moved with his family from Paris to Provence, where they stayed with his cousin Alexandre Maubert at Grasse. When the Terror was over they returned to Paris, and in 1792 Alexandre-Evariste entered the studio of David, by then the most important and influential painter in France. David had a great respect for Fragonard, and it was perhaps out of gratitude for an earlier kindness that he recommended the elderly artist for membership of the newly-formed Museum des Arts, an administrative post which saved Fragonard from penury and entitled him to free lodgings in the Louvre. These he was compelled to leave in 1805 when the Emperor Napoleon decided to take over more of the Louvre for use as a museum. Fragonard then went to live in the Palais Royal. He died on 22 August 1806 on his way home from a restaurant, an obscure and forgotten figure, a relic of the vanished society which had idolized him. His art was then totally eclipsed for at least the next thirty years. After 1830 certain minor Romantics like Devéria began to make engravings after Fragonard, but it was only about the middle of the nineteenth century that Daumier and collectors like La Caze and Walferdin took a serious interest in him. The Goncourt brothers's study of 1865, supported by Portalis's monumental book of 1889, further confirmed Fragonard's growing reputation, but in our own century this has again ebbed and is by no means universally recognized. Fragonard remains a great unknown figure among French artists.

The reasons for this comparative neglect are complex. In the first place his art is widely dispersed all over the world, and it is not easy to form a complete picture of it from any single collection. His greatest work, *The Fête at Saint-Cloud*, has very rarely been on public exhibition. Another reason may be the egalitarian bias of our own age which has brought about a shift of interest away from the Ancien Régime, with which Fragonard was associated, towards the more democratic art of the Revolution. A third and perhaps still more decisive reason why Fragonard is given short shrift is the tendency among some art historians to identify painters with a particular cultural or historical phase of civilization. Thus, Watteau becomes the painter of the Régence, Boucher of the reign of Louis XV and the Pompadour style, Chardin is the representative of the humble artisan class, Greuze the leader of the moral reaction away from Rococo frivolity and David the hero of the Revolution. In this over-simplified scheme there is no place for such a complex and variable artist as Fragonard, who cannot readily be linked to any such movement or intellectual trend and who defies all attempts at categorization.

The truth is, perhaps, that he touched on all these different strands without being bound by any one of them. Sometimes his work harks back nostalgically to the Régence of Watteau, at others it seems to herald the mood of the Revolution. This inconsistency of style and lack of a single-minded purpose make him the outstanding individual among eighteenth-century painters. He was the supreme improvisor, dashing off sketches often without much thought for the finished product; although when he worked for patrons he painted to satisfy himself as much as his client. In his later works especially he seemed to strive after an entirely free form of self-expression, independent of the accepted genres and categories of his time. Unlike Chardin and Boucher, who belonged more to the old-fashioned type of artist-craftsman, Fragonard was not content to execute technically perfect works within a narrow range. His restless impatience, lack of concern with form and sheer delight in creation, even at the cost of carelessness, made Fragonard, in one sense, more typical of the nineteenth-than the eighteenth-century painter. And yet, in another sense, he seems the typical artist of the Ancien Régime. His art epitomizes a period of aristocratic leisure, romps in parks, swings, games of blind man's buff, harlequinades, idle noblemen with nothing better

to do than loll in the grass, unreflecting people enjoying themselves in happy oblivion of the harsher aspects of eighteenth-century life. Here Fragonard takes us back to Watteau's enchanted magic fantasies, such as the *Champs Elysées* in the Wallace Collection, w: small groups of people nonchalantly strolling in park opening out into vast blue perspectives.

Fragonard, however, has none of Watteau's melancholy and sense of decadence, unless we write this off as the invention of Nerval, Gautier and other Romantic critics, all too ready to discover precocious symptoms of the *mal du siècle* in the eighteenth century. This difference of outlook, passive and diffident in Watteau, robust and optimistic in Fragonard, may partly be ascribed to temperament and background, but also to a change in the climate of the times. People in late eighteenth-century France did not live in fear and trembling of the impending Revolution. Until 1780 at least it was a period of hopeful optimism, when the Encyclopédistes believed that they could put the world to rights if only they could persuade a well-intentioned king and government to carry out some necessary practical reforms. Even the more enlightened members of the nobility, swayed by the fashionable new ideas, were prepared to sacrifice some of their privileges, and the art and literature of the time are full of well-born people who willingly, perhaps condescendingly, abdicate their social status to lead simpler lives. There was a greater mixing of the social classes than ever before in France, certainly than in the seventeenth century, dominated by hierarchy and etiquette. This new informality and move away from court and urban life lies at the core of Fragonard's art. Thus we find in his pictures not only privileged aristocrats, as in Watteau's, but also plebeians enjoying a day out in the country, stable lads and village schoolmistresses, young girls of all classes of society, reading, writing, making love. We already seem to be in a world where the barriers have broken down, indeed Fragonard seems blithely unaware of their existence. In this sense his art is far more representative of life in eighteenth-century France than either Watteau or Boucher, even though the Goncourts saw Boucher as the typical painter, the 'incarnation' of the period, while being far less confined to a particular aspect of taste. By his sheer lack of inhibition, by his range and variety, his willingness to tackle anything which came his way, Fragonard managed to create a microcosm of the Ancien Régime, with all its contradictions and growing tensions well buried below a surface of hedonistic enjoyment.

1 *Woman Gathering Grapes*
formerly CHICAGO, collection of Mrs. Fowler McCormick.
c. 1750. Oil on canvas in gilt frame 149 × 83 cm.

This is an early work in which Boucher's influence is
manifest in the smoky blue-green tonality and the
picturesque rusticity, with random objects scattered in
the foreground and the characteristic dovecot in the
background. A comparison between the *Woman
Gathering Grapes* and the great park scenes of the 1770s
shows how far Fragonard outgrew Boucher's genteel
pastorals, although there is already a hint of Fragonard's
individual style in the animal zest of the two children
and the sidelong glance of the young woman. Paintings
like this were perhaps intended to be reproduced by the
Gobelins tapestry factory.

(page 17)
I *The Washerwomen*
AMIENS, Musée de Picardie. c. 1761–62. Oil on canvas 47 × 65 cm.

Given to the Musée de Picardie by the Lavalard brothers in 1894, *The Washerwomen* belongs, with
The Waterfall at Tivoli (Plate 11) and *The Gardens of the Villa d'Este* (Plate 12), to the group of
paintings produced in the years immediately following Fragonard's return from Italy to Paris in
1761. The site of this particular painting is not known, but the crumbling decay of the steps and
statues gives a good idea of the state of the Villa d'Este gardens when Fragonard stayed there in
1760. Hubert Robert and Fragonard both liked to paint this vision of classical Antiquity, decaying
but still in daily use by the washerwomen who strung out their lines from one statue to another.

(left)
II *The Swing*
LONDON, Wallace Collection. 1766 or 1767. Oil on canvas 81 × 65 cm.

In 1859 the duc de Morny offered *The Swing* to the Louvre, which turned it down, and it was
bought instead by Lord Hertford in 1865 for 30,000 francs. The painting was commissioned from
Fragonard by the baron de Saint-Julien in 1766. In his *Journal*, Collé relates how Saint-Julien first
approached the history painter Doyen for a picture of his mistress on a swing, with a bishop
standing by and the baron lolling in the grass in a strategic position from which to observe the
girl's legs. Doyen indignantly refused the commission and handed it on to Fragonard, who carried
out the picture according to Saint-Julien's wishes, except that in place of the bishop he put an
acquiescent husband on the right in the background, giving a helpful push to the swing. The statue
of Cupid on the left is probably Falconet's *L'Amour menaçant* from the Louvre, exhibited at the
Salon of 1757.

2 *The Four Seasons*

Winter. LOS ANGELES, Los Angeles County Museum of Art (William Randolph Hearst collection).
Oil on canvas 80 × 164 cm.

The Four Seasons were executed around 1750–52, still very much in the manner of Boucher, as
decorative overdoors for the salon of the Hôtel Matignon, 57 rue de Varenne, Paris. They were
probably commissioned by the owner, the duc de Valentinois, formerly Jacques I, Prince of
Monaco, who died in 1751. The building then became the Austro-Hungarian Embassy and is now
the Présidence du Conseil. Only three of the original panels are still in place. The fourth, *Winter*,
was removed and sold when the house was remodelled soon after its construction.

(overleaf)
3 *Blind Man's Buff*
TOLEDO (OHIO), Toledo Museum of Art (gift of Edward Drummond Libbey). c. 1750–52. Oil on canvas 114 × 90 cm.

Clearly painted as a pendant to *The See-saw* (Plate 4), which it resembles in style and size, *Blind Man's Buff* was possibly commissioned by the baron de Saint-Julien. An early work, the lively handling of the paint and cheerful disorder of the composition betray the influence of Boucher. In fact, both this painting and *The See-saw* were thought by the Goncourts (*L'Art du dix-huitième siècle*, vol. III, p. 331) to be simple pastiches of Boucher. There are two other versions, neither accepted by Wildenstein, one large panel in the Petit Palais, Paris, and another in the Walters collection, New York.

22

4 *The See-saw*

AMSTERDAM, Stichting Collectie Thyssen-Bornemisza. c. 1750–52. Oil on canvas 114 × 90 cm.

The pendant to the Toledo *Blind Man's Buff* (Plate 3).

5 *Jeroboam Sacrificing to the Golden Calf*
PARIS, École des Beaux-Arts. Oil on canvas 115 × 145 cm.

Presented by Fragonard for the Prix de Rome in 1752, the painting was awarded first prize. The second prize was won by Gabriel de Saint-Aubin. This work is Fragonard's first essay in the grand historical style, which he quickly mastered, a remarkable achievement for a young artist of twenty who had received no academic training. A clear statement of the painter's adherence to the Rubéniste, colourist faction which triumphed in French art during the first half of the eighteenth century, it contains a strong reference to Poussiniste classicism in the superbly modelled incense bearers in the foreground. The Goncourt brothers wrote of this picture that 'The animated groups, the lively draperies, the cloudy splendour of the architecture, the whites and reds, the colour of a more vaporous, wash-like de Troy promise a great deal for Fragonard's future as a painter.'

6 *The Storm*
PARIS, Musée du Louvre. c. 1759. Oil on canvas 73 × 97 cm.

Passing from the La Caze collection to the Louvre in 1869, *The Storm* is the subject of a preparatory
study in the Art Institute of Chicago, executed in brown chalk, signed and dated: Rome, 1759.
The painting is probably derived from a landscape drawing by Castiglione (1610–65), perhaps
The Pastoral Journey (Louvre, Cabinet des Dessins), and yet there is something unmistakably
Northern in the ominous, lowering sky which clearly anticipates English landscapes by Crome and
Gainsborough. Fragonard could transform an everyday episode from a farmer's life into a
momentous drama.

7   *Annette and Lubin*
WORCESTER (MASSACHUSETTS), Worcester Art Museum (gift of Theodore and Mary Ellis). c. 1762–65.
Oil on canvas 65 × 80 cm.

Fragonard painted several versions of this theme, including *Annette at the Age of Twenty* in the Galleria
Nazionale d'Arte Antica, Rome, and its pendant, *Annette at the Age of Fifteen*, now lost, but engraved in
1772 by F. F. Godefroy. Godefroy's engravings after both these pictures were announced in the *Mercure
de France* in 1772. The subject is taken from the story 'Annette et Lubin' in the *Contes Moraux* (1762),
a collection of short pastoral idylls by Marmontel, and relates the love of two cousins, innocent
young shepherds; Annette, finding herself pregnant, is told by an officious local judge that she has
committed mortal sin and forbidden to marry her lover. Fortunately a benevolent squire comes to
the rescue and obtains a special dispensation from the Pope for the cousins to marry. J. F. Marmontel
(1723–99) was one of the most popular writers of his day, a prominent literary critic, friend of
Mme. Geoffrin and editor of the *Mercure de France*.

8 *The Giant Cypresses at the Villa d'Este at Tivoli*
BESANÇON, Musée des Beaux-Arts. Brown chalk on white
paper 47·8 × 35·4 cm.

Collections: Saint-Non; P. A. Pâris; Bibliothèque
Municipale de Besançon.

9 *The Entrance to Fontanone at the Villa d'Este*
BESANÇON, Musée des Beaux-Arts. Brown chalk on white
paper 48·8 × 36·1 cm.

Collections: Saint-Non; P. A. Pâris; Bibliothèque
Municipale de Besançon.

10 *View of the Coast near Genoa*
BESANÇON, Musée des Beaux-Arts. Pen and bistre wash on
pencil sketch 35·3 × 46·2 cm. Signed on the left at the
bottom: Fragonard.

It probably was executed during Fragonard's second visit
to Italy in 1774. Collections: Bergeret (?); P. A. Pâris;
Bibliothèque Municipale de Besançon.

12 *The Gardens of the Villa d'Este at Tivoli*
LONDON, Wallace Collection. c. 1762. Oil on canvas 38 × 46 cm.

Engraved by Fragonard in 1763 and by Saint-Non, the painting was exhibited at the Salon de la Correspondance in July 1785. One of the views of Italy painted soon after Fragonard's return to Paris in 1761, it is similar in style and subject to *The Waterfall at Tivoli* and *The Washerwomen*. The wonderful precision and firmness of these paintings was the result of many preparatory drawings which Fragonard made in the summer of 1760 at the Villa d'Este in the company of Hubert Robert and the abbé de Saint-Non.

(left)
11 *The Waterfall at Tivoli*
PARIS, Musée du Louvre. c. 1761–62. Oil on canvas 73 × 60 cm.

Given to the Louvre from the La Caze collection in 1869, this picture is probably the one mentioned in the inventory of Saint-Non's collection (1792) as 'Paysage avec cascades et figures, vue de Tivoli'. Formerly attributed to Hubert Robert, it was restored to Fragonard's authorship by Charles Sterling. Despite a close affinity between the two artists, detailed inspection shows that Fragonard's technique is much bolder and less meticulous than Robert's, in, for example, the washing on the line, which consists solely of small patches of white paint. The sheer technical mastery and spontaneity, disciplined by compositional rigour, of this and the other Italian landscapes have rarely been surpassed.

29

13 *Rinaldo in the Gardens of Armida*
PARIS, collection of M. Arthur Veil-Picard. 1761–65. Oil on canvas 72 × 91 cm.

The subject is from Canto XVI of Tasso's epic poem, *Gerusalemme Liberata*. It is unlikely, however, that Fragonard had read Tasso in the original text, and his immediate source would seem to be the opera *Renaud et Armide* by Lulli and Quinault, performed in Paris in 1761 and 1764. The Rococo theatricality of this picture invites comparison with Tiepolo, who also painted a version of Rinaldo and Armida, 1750–55 (Chicago, Art Institute); but whereas Tiepolo showed Rinaldo seated by Armida's side, Fragonard, with his love of impetuous movement, chose the moment when Rinaldo bursts on to the stage like an operatic hero. Both this picture and its pendant (Plate 14) show that Fragonard had already developed his own individualistic, highly coloured idiom.

14 *Rinaldo in the Enchanted Forest*
PARIS, collection of M. Arthur Veil-Picard. 1761–65. Oil on canvas 72 × 91 cm.

The pendant to Plate 13.

16 *Plutarch's Dream*
ROUEN, Musée des Beaux-Arts. Oil sketch on canvas 24 × 31 cm.

Although catalogued by Wildenstein (cat. no. 85) as an early work, *Plutarch's Dream* would seem to belong to the group of literary compositions of the 1770s. The historian is seen reading a copy of his *Lives of Famous Men*. The painting was acquired by the Musée des Beaux-Arts, Rouen, in 1818. There is a drawing of the same subject in the collection of Mme. Fernand Halphen, Paris (Ananoff, fig. 154, cat. no. 453).

(left)
15 *The Warrior's Dream* or *Le Songe d'amour*
PARIS, Musée du Louvre. c. 1761–65. Oil on canvas 60 × 50 cm.

Given to the Louvre in 1915 by the baron de Schlichting, *The Warrior's Dream* is one of Fragonard's allegorical, quasi-literary compositions, comparable to *The Sculptor's Vision* and *Plutarch's Dream* (Plate 16). It shows a warrior, dressed in a toga, asleep; Venus appears to him in a dream. It is not clear whether this picture has a precise source, but its mixture of dream, war and erotic fantasy relates it to the world of Ariosto and Cervantes; or perhaps it is only a variant on the theme of Mars disarmed by Venus, popular with many French artists around 1760.

17 *Coresus Sacrificing Himself to Save Callirhoe*
PARIS, Musée du Louvre. Oil on canvas 309 × 406 cm.

Exhibited at the Salon of 1765, Fragonard's 'morceau de réception' won him a place in the
Academy in the same year. The painting was commissioned by Marigny, the Directeur des
Bâtiments, who also promised to have the work reproduced in tapestry by the Gobelins factory,
work which was never carried out. The *Coresus* was a great success with all the critics of the Salon,
including Cochin, Diderot and Fréron, but because the royal finances were precarious at this time
Fragonard was poorly paid and only belatedly, in three successive instalments in 1765, 1766 and
1773. This experience of royal patronage, as well as his own natural bent, may have put
Fragonard off further attempts at history painting. The action represented is a typical case of heroic
sacrifice popular in the second half of the eighteenth century. The Princess Callirhoe has been
designated as a victim to ward off a plague; Coresus, the high priest of Bacchus who is in love with
her, offers himself for sacrifice in her place and is about to stab himself in the chest. The
immediate source for the picture is not certain, but there are two plausible alternatives, the
tragedy by La Fosse, *Corésus et Callirhoë*, first performed in 1703, or the opera by Le Roy,
Callirhoë, first performed in 1712 and frequently later. There is no doubt about the operatic,
melodramatic character of the work, which anticipates the most extreme sadistic horrors of
Romanticism, Delacroix's *Death of Sardanapalus* for instance. Fragonard's picture was probably
intended not as an *exemplum virtutis* but as an indictment of religious fanaticism. By an ironic
accident it was criticized by the Revolutionary régime in 1793 on the grounds that it 'evoked
superstitious ideas'. There are numerous preparatory sketches for *Coresus*, and later replicas include
versions at Madrid and Angers.

III *Imaginary Figure* or *The Actor*

PARIS, Musée du Louvre. 1769. Oil on canvas 80 × 65 cm. On the back of the picture an inscription: Portrait de l'abbé de Saint-Non, peint par Fragonard, en une heure de temps.

This is another in the series of fancy portraits showing the abbé de Saint-Non masquerading as an actor or literary hero.

18 *The Bathers*
PARIS, Musée du Louvre. c. 1765–70. Oil on canvas 65 × 81 cm.

A whirl of naked bodies and yellow-green foliage; the unity of texture of this painting, in which water and clouds seem to be made of the same substance, make one of Fragonard's most modern compositions, a direct precursor of Renoir's *Grandes baigneuses* of 1887. The painting was in the La Caze collection and was given to the Louvre in 1869.

(left)
IV *The Pursuit* (Plate 34b)
NEW YORK, Frick Collection

(*above*)
19 *The Three Trees*
formerly PARIS, collection of M. Cailleux. c. 1762–65. Oil on canvas 34 × 24 cm.

This most haunting of a series of landscapes in the Northern manner appeared in the Walferdin sale, April 1880. It is strongly reminiscent of Ruysdael, with brooding skies and melancholy trees, and the question has been raised whether Fragonard visited Holland, although there is no conclusive evidence to show that he did. It is certain, however, that he had ample opportunity to study the Dutch masters who figured prominently in Parisian collections in the eighteenth century. Fragonard perhaps wished to latch on to the new trend away from the Italian masters towards Dutch 'cabinet pictures' which became increasingly popular with collectors during the century. See the article by J. Wilhelm, 'Fragonard as a Painter of Realistic Landscapes', *Art Quarterly*, Autumn 1948.

(*opposite, top*)
20 *Landscape with a Large Stretch of Greensward*
GRASSE, Musée Fragonard. c. 1762–65. Oil on canvas 63 × 91 cm.

Formerly in the Louvre, but sent to Grasse in 1887, the painting is similar in composition to *Annette and Lubin* (Plate 7), a pastoral landscape with sheep and a hillock slightly off centre.

(*opposite, bottom*)
21 *The Glade* or *Landscape with a Horseman*
DETROIT, Detroit Institute of Arts. Oil on canvas 37 × 44 cm.

Given to the Detroit Institute by Mr. and Mrs. Edgar B. Whitcomb in 1948.

22 *The Music Lesson*

PARIS, Musée du Louvre. c. 1765–72. Oil on canvas 110 × 120 cm.

One of Fragonard's most popular paintings, and rightly so; the fond look of the boy turning over the pages and the characteristically awkward stance of the young girl practising the keyboard are perfectly observed. The spectator can almost hear the fumbling notes of a piece by Rameau or Couperin. Technically too the picture is masterly. With the shimmering satin of the girl's dress and the still-life on the right, Fragonard combined the best of both his masters, Boucher and Chardin, in this perfect evocation of domestic life in eighteenth-century France. *The Music Lesson* was in the Walferdin collection, was acquired by the French government in 1849, but was removed from the Louvre in 1865 to decorate the house of Napoleon III's private secretary. It was restored to the Louvre in 1870.

(*left*)
23 *Study of a Man Seated on a Stone*
PARIS, Musée de Louvre, Cabinet des Dessins. Bistre wash 36·3 × 28·6 cm. Inscribed: Rome, 1774.

This remarkable drawing shows that Fragonard's capacity for the realistic observation of ordinary people was just as strong as Chardin's or Watteau's.

(*right*)
24 *Young Girl and her Little Sister*
BESANÇON, Musée des Beaux-Arts. Chalk drawing on white paper 44·5 × 28·6 cm. Signed on the left: Fragonard.

Formerly in the P. A. Pâris collection, this sketch was executed in 1774, during the artist's second journey to Italy.

25 *Music: Portrait of M. de la Bretèche*
PARIS, Musée du Louvre. 1769. Oil on canvas 80 × 65 cm. Signed in left-hand corner: Frago 1769; on back: Portrait de la Bretèche, peint par Fragonard, en une heure de temps.

The painting passed from the La Caze collection as a gift to the Louvre in 1869. M. de la Bretèche was the brother of the abbé de Saint-Non, and this is the first and best-known of a series of about fifteen 'figures de fantaisie', or fancy portraits, all of them painted with extraordinary virtuosity, reducing the painted surface to long strands of colour. The sitters, who included Diderot and Sophie Guimard (Plate 28), all bear a family likeness, indeed they seem almost to be the same person in different disguises. There are two others of the duc d'Harcourt and the duc de Beuvron (French private collection), another known as *The Actor* (Rothschild collection), the mysterious *Warrior* (Plate 27) and *Saint-Non in Spanish Costume* (Plate 29). The remaining seven are in the Louvre, including *Song* (or *Study*), *The Actor* (Plate III) and *Inspiration* (Plate 26), which, with the portrait of *M. de la Bretèche*, may be representations of the four arts, music, drama, poetry and song. M. Thuillier has pointed out the affinity between Fragonard's portraits and Rubens's cycle of *The Life of Marie de'Medici*, especially in the style of costume and dress. Fragonard's portraits are also strongly sculptural in feeling. The artist handled paint like a sculptor modelling clay, and there is often a resemblance between his portraits, detached on their pedestals in half-profile, and busts by Houdon.

26 *Inspiration*
PARIS, Musée du Louvre. 1769. Oil on canvas 80 × 60 cm.

Another of the 'figures de fantaisie', *Inspiration* probably represents the art of poetry. The picture
has been thought to be a portrait of the abbé de Saint-Non, and the sitter undoubtedly has similar
features to the abbé on the basis of Saint-Non's portrait in Spanish costume (Plate 29). Again, the
precise identity of the subject is not of great importance, for what Fragonard created is an
impersonal depiction of the act of literary creation.

27 *The Warrior*
WILLIAMSTOWN, Sterling and Francine Clark Art Institute. c. 1770–73. Oil on canvas 81 × 65 cm.

Bought in 1774 by Stanislas Poniatowski, King of Poland, *The Warrior* then belonged to Count Potocki and was recently acquired by the Sterling and Francine Clark Institute (See C. Sterling, *An Unknown Masterpiece by Fragonard*. Special publication of the Clark Institute, 1964). It is one of the most striking and still least known of Fragonard's portraits. The sitter is wearing a bright yellow tunic, white ruff and black cape lined with red, with the handle of his sword thrust aggressively forward. Whoever he is, and his identity is unknown, this mysterious figure has the hard-bitten, deeply cut features of some heroic desperado with something of the stoical disillusionment present in Daumier's portraits.

28 *Portrait of Sophie Guimard*
PARIS, Musée du Louvre. c. 1770. Oil on canvas 81 × 64 cm.

Formerly in the Watel-Dehaynin collection, the painting was acquired by the Louvre in 1972. The sitter is probably the ballet dancer, Mlle. Guimard. Since Fragonard is known to have worked for her between 1770 and 1773 this portrait probably dates from those years; it is also clearly related in style to the fancy portraits in the Louvre of the same period. The colour scheme of this picture was a favourite one with Fragonard in his maturity, a brown ground with a few streaks of red, contrasted with the white sleeves, ruff and head-dress.

29 *The abbé de Saint-Non in Spanish Costume*
BARCELONA, Museum of Modern Art. Oil on canvas 94 × 74 cm.

Another of the 'figures de fantaisie' painted in or around 1769, this portrait depicts the abbé de Saint-Non dressed up in a red tunic, white ruff and black plumed hat, holding a huge sword in one hand and, in the other, the reins of his horse tethered behind him. The most likely explanation is that Saint-Non is masquerading as Don Quixote. Fragonard is known to have planned a series of illustrations for Cervantes's novel (Plates 48a, b).

(*above*)
30 *The Longed-for Moment* or *The Happy Lovers*
PARIS, collection of M. Arthur Veil-Picard. 1765–70. Oil on canvas 55 × 65 cm.

The Longed-for Moment appeared in the Walferdin sale, 1880. In the appendix to their study of Fragonard (1865) the Goncourts reported that at 'M. Walferdin's house there were some of the Master's finest pictures; a young man kissing a woman lying on a bed. . . .'

(*opposite, top*)
31 *The Stolen Shirt*
PARIS, Musée du Louvre. c. 1765–72. Oil on oval canvas 36 × 43 cm.

Originally in the La Caze collection, *The Stolen Shirt* was given to the Louvre in 1869. It is the pendant to *All Ablaze* (Plate 32) and was sketched by Gabriel de Saint-Aubin in the margin of the Gros sale catalogue of 1778. One of the artist's most accomplished erotic pieces, it is as neat and deft as the action of the cupid, shown undressing the girl on the bed.

(*opposite, bottom*)
32 *All Ablaze* or *Le Feu aux poudres*
PARIS, Musée du Louvre. c. 1765–72. Oil on oval canvas 37 × 43 cm.

As with its pendant (Plate 31), *All Ablaze* was sketched by Saint-Aubin in the margin of the Gros sale catalogue of 1778. Here the cupid is seen setting the girl ablaze with passion with his torch.

47

33 *The Bed with Cupids*
BESANÇON, Musée des Beaux-Arts. Pen and ink wash with
watercolour on white paper 45·7 × 30·4 cm. On the left,
at the bottom, in Pâris's hand: Fragonard.

Collections: P. A. Pâris; Bibliothèque Municipale de
Besançon.

34 *The Pursuit of Love*
NEW YORK, Frick Collection. 1770–72.

Storming the Citadel. Oil on canvas 318 × 224 cm.

The Pursuit. Oil on canvas 318 × 215 cm.

The Declaration of Love. Oil on canvas 318 × 215 cm.

The Lover Crowned. Oil on canvas 318 × 243 cm.

Abandonment (oil on canvas 318 × 196 cm.), the fifth panel, may also belong to the series but was
probably painted later, around 1790, to decorate the dining room of Fragonard's cousin Maubert
at Grasse.

The series of decorative panels known as *The Pursuit of Love* was originally commissioned, probably
in 1770, by Mme. du Barry for her new pavilion at Louveciennes. The original château at
Louveciennes was a modest affair and had been given to Mme. du Barry in July 1769. In 1770 she
ordered the architect Ledoux to build her a new house in the latest Neo-Classical style, with a
severe façade and recessed Ionic columns. This was the building for which Fragonard's panels were
intended, and he probably started work in the autumn of 1771. On 22 July 1772 Bachaumont
reported in his *Mémoires secrets* that 'the curious are going in large numbers to Louveciennes to see
the pavilion of Madame du Barry; but not everyone who wants can enter, and it is only by a
special favour that one can penetrate this sanctuary of pleasure.' Then in 1773, for unstated reasons,
she returned Fragonard's panels and paid him 18,000 livres compensation. She replaced his work
with paintings by Joseph-Marie Vien, whose pseudo-Greek style had become fashionable in smart
society. The reason for Mme. du Barry's sudden change of mind has recently been most
satisfactorily explained in an article by F. M. Biebel (*Gazette des Beaux-Arts*, October 1960), namely
that Fragonard's Rococo style was already considered out of date by the court circle who preferred
the attenuated Greek style of Vien. Thus Fragonard was overtaken by the rapidly changing taste of
late eighteenth-century France. After the rejection of his panels Fragonard rolled them up, put
them in store and in 1790 took them to Grasse, where he used them to decorate his cousin's house.
He also painted other panels there of similar size and theme, including, probably, the
Abandonment. In 1806 Malvilan, who then owned the house, offered to sell them to Fragonard's son
Alexandre-Evariste; he declined the offer. In 1857 they were offered to the Louvre, which, for lack
of funds, was unable to buy them. Finally in 1898 Malvilan's heirs sold the canvases to
Wertheimer in London, who sold them to Pierpont Morgan. They were acquired by Frick in 1915.

34a *The Pursuit of Love: Storming the Citadel*

34b *The Pursuit of Love : The Pursuit*

34c *The Pursuit of Love: The Declaration of Love*

34d *The Pursuit of Love: The Lover Crowned*

V *The Swing* (Plate 36)
WASHINGTON, National Gallery of Art (Samuel H. Kress collection)

(left)
VI *The Love Letter* or *Le Billet Doux*
NEW YORK, Metropolitan Museum of Art (Jules S. Bache collection). c. 1775. Oil on canvas 83 × 67 cm.

Formerly in the Feuillet de Conches collection, *The Love Letter* is one of Fragonard's most bewitching studies of women. The sitter is tall and slender, with an oval-shaped face, unlike some of his nudes who tend to be chubby and featureless. She is sitting at a desk, half-turned towards the spectator, clasping a bunch of flowers with a love message which cannot be deciphered with certainty. The exact wording of the message is less important than Fragonard's exceptional ability to portray the state of emotional anticipation in a woman. When André Gide saw this painting at an exhibition of the Crosnier collection in 1905 he found it 'delightful' but considered Fragonard's reputation to be exaggerated (Gide, *Journal*, ed. Pléiade, p. 186).

35 *Blind Man's Buff*
WASHINGTON, National Gallery of Art (Samuel H. Kress collection). c. 1775. Oil on canvas
215 × 197 cm.

The pendant to *The Swing* (Plate 36) and one of the great decorative panels related to *The Fête at Saint-Cloud* and *The Fête at Rambouillet, Blind Man's Buff* was painted probably soon after Fragonard's return from his second journey to Italy in 1774. This and the other Washington picture were probably in the collection of Saint-Non, catalogued as 'Un paysage fait en Italie avec figures, sujet de la Main Chaude'. As in all the works of Fragonard's maturity, his treatment of the landscape, whether actually in France or in Italy, is unmistakably Italianate, with tall cypresses, fountains and statues. On the left a small group of young people is playing blind man's buff, while on the right others are preparing a picnic. The whole work has something of the atmosphere of a Venetian carnival seen by Tiepolo.

36 *The Swing*
WASHINGTON, National Gallery of Art (Samuel H. Kress collection). c. 1775. Oil on canvas 215 × 186 cm.

The pendant to *Blind Man's Buff*, *The Swing* may be the painting mentioned in the 1791 inventory of the collection of the abbé de Saint-Non as a 'Paysage d'Italie par Fragonard avec figures, sujet d'une balançoire'. One of Fragonard's most beautiful creations, the scene is set, perhaps, in a royal park just outside Paris, although the peculiar shape of the mountain in the distance suggests a more exotic Mediterranean landscape. Like Watteau, Fragonard had a fine feeling for the grouping of figures, and the couple on the park bench, especially the woman looking through the telescope, is one of the most perfect details in his work.

36a *The Swing* (detail)

37 *Blind Man's Buff*
PARIS, Musée du Louvre. c. 1775. Oil on canvas 36 × 45 cm.

Acquired by the Louvre in 1926 from M. Alexandre Poliakoff, this version of *Blind Man's Buff* was mentioned in the catalogue of the Gros sale in 1778 with its description as 'A landscape of a fresh, rustic kind set in the neighbourhood of Meudon. In the foreground and on a terrace one can see several groups of girls and boys, some of whom are preparing to play blind man's buff and others, next to a table, are busy with different kinds of games.' Underneath this caption is a sketch by Gabriel de Saint-Aubin. This painting is perhaps the one in which Fragonard came closest in spirit to Watteau.

38 *Landscape with Two Figures*
BARNARD CASTLE, Bowes Museum. c. 1770–75. Oil on canvas 27 × 20 cm.

This tiny, little-known landscape shows Fragonard in a mood clearly influenced by Watteau.
Related in style to the Washington *Swing* and *Blind Man's Buff* and probably dating from the
1770s, it appeared in Agnew's exhibition (1952) of pictures from the Bowes Museum (no. 45) and
at the Royal Academy in the 1954 European Masters of the Eighteenth Century exhibition
(no. 204).

39 Illustrations for Ariosto's 'Orlando Furioso'

Many of these were published in the volume *Fragonard: Drawings for Ariosto* (London, 1946), with essays by Elizabeth Mongan, Philip Hofer and Jean Seznec. It contains 137 drawings by Fragonard for an edition of the *Orlando Furioso* which never appeared. Their first owner was Hippolyte Walferdin and they were bought at his sale in 1880 by Louis Roederer. The illustrations were dispersed in 1923 and are now scattered in various collections in America, Paris and Berlin. About 150 drawings for Ariosto are now known to exist, and others may be in circulation. All the drawings in this series correspond closely to specific episodes in Ariosto's epic poem. Although Fragonard displays abundant verve and creative power, sometimes bordering on the fantastic, he sticks closely to the text. In fact, as Jean Seznec remarks, the natural affinity between Fragonard and Ariosto and their common Mediterranean heritage made the one the ideal interpreter of the other. These wonderfully sparse and luminous drawings have all the clarity of a poetic image; those reproduced (Plates 39b–f) have been grouped to illustrate the episode of Roger freeing Angelica, a theme which also held a strong appeal for Ingres and the Romantics.

39a Ariosto Inspired by Love and Folly
BESANÇON, Musée des Beaux-Arts. Bistre wash on black chalk 3·3 × 4·5 cm.

Formerly in the collection of Pierre-Adrien Pâris, the drawing entered the Musée des Beaux-Arts, Besançon, in 1843. Possibly intended as the frontispiece for the edition of *Orlando Furioso*, it shows Ariosto composing his poem under the twin inspiration of Love and Folly.

61

39b *Ruggiero Blinds the Orc*
formerly PHILADELPHIA, collection of A. S. W. Rosenbach. Black chalk and bistre wash.

From Canto X of *Orlando Furioso*.

39c *Ruggiero Unchains Angelica*
formerly PHILADELPHIA, collection of A. S. W. Rosenbach. Black chalk and bistre wash.
From Canto X of *Orlando Furioso*.

39d *Ruggiero Divests Himself of his Armour*
PHILADELPHIA, the Rosenbach Foundation. Black chalk and bistre wash.

From Cantos X and XI of *Orlando Furioso*.

39e *Ruggiero Seeks Angelica who Has Made Herself Invisible*
formerly PHILADELPHIA, collection of A. S. W. Rosenbach. Black chalk and bistre wash.

From Canto XI of *Orlando Furioso*.

39f *Ruggiero Despairs at the Loss of Angelica*
PHILADELPHIA, the Rosenbach Foundation. Black chalk and bistre wash.

From Canto XI of *Orlando Furioso*.

(left)
40 *The Charlatans*
ZÜRICH, collection of Emile Bührle. c. 1770–75. Oil on canvas 49 × 38 cm.

A study for the left hand part of *The Fête at Saint-Cloud* (Plate 42), this large sketch differs from the final version in that the figure on the rostrum in front of the placard is holding a sword and wears a red cap. He is probably one of the figures from the Commedia dell'Arte. The beribboned placard seems to be a kind of programme of events illustrated by a strip cartoon round the border. Whether actor, quack-doctor or simple mountebank, his function is to entertain the crowd thronged around the stall. This was just one of the amusements provided for the Parisian people.

(right)
41 *The Toy Seller*
ZÜRICH, collection of Emile Bührle. c. 1770–75. Oil on canvas 40 × 33 cm.

This is a study for the central part of *The Fête at Saint-Cloud*. The toys for sale can be seen hanging from the roof of the booth, glowing with luminous colour.

(overleaf)
42 *The Fête at Saint-Cloud*
PARIS, Banque de France. Oil on canvas 216 × 335 cm.

Probably painted in 1775 for the duc de Penthièvre to decorate his apartment in the Hôtel de Toulouse, since 1808 the Banque de France, *The Fête at Saint-Cloud* now hangs in the Governor's dining room, which it has rarely left. It was recently exhibited in Paris in 1975. See G. Wildenstein, 'La Fête de Saint-Cloud et Fragonard', *Gazette des Beaux-Arts*, January 1960.

42a *The Fête at Saint-Cloud* (detail)

VII *The Fête at Saint-Cloud* (detail)

43 *The Fête at Rambouillet* (detail of Plate VIII)

(left)
VIII *The Fête at Rambouillet* or *The Island of Love*
LISBON, Calouste Gulbenkian Foundation. c. 1775. Oil on canvas 72 × 91 cm.

Exhibited at the Salon de la Correspondance in 1782 under the title 'Une grotte ornée
d'architectures avec figures', *The Fête at Rambouillet* is another key work in Fragonard's oeuvre and
one which poses several problems. First, the dating. A stylistic comparison with *The Fête at
Saint-Cloud* would indicate that it was painted around 1775. The setting is almost certainly the
Island of Rocks in the park of the Château de Rambouillet, which then belonged to the duc de
Penthièvre; it is reasonable to suppose that it was he who commissioned the painting. The subject
is not altogether plain, but it may be one of the parties held in honour of Louis XV's frequent
visits to Rambouillet. The painting seems like a remote echo of Watteau's *Embarkation for Cythera*,
some refined form of aristocratic entertainment. A group of elegant men and women can be seen
descending the steps from the terrace, while others are setting out in a boat into a rocky, swirling
lake. The foaming waters and gnarled tree in the centre, reminiscent of Salvator Rosa, point to an
obvious contrast between the security of the group of spectators behind the balustrade and the
elemental forces of nature. There is, however, a theatrical unreality about the picture which
suggests that fashionable society is merely enjoying the thrill of exposing itself to mild dangers.

(left)
44 *The Souvenir* or *A Lady Carving her Name*
LONDON, Wallace Collection. c. 1775–80. Oil on canvas 25 × 19 cm. Signed on the edge of the bench: Fragonard.

Bought in 1865 by Lord Hertford for 35,000 francs at the Morny sale, the painting is related to a lost composition, recorded by Wildenstein (cat. no. 389) and entitled variously *A Lady Carving her Name* or *Angelica Writing the Name of Medor on a Tree* or *Fair Julia*. *The Souvenir* shows Fragonard at his most wistful and poetic, a mood which Corot later recaptured in works like the *Souvenir de Mortefontaine.*

(right)
45 *Portrait of Marguerite Gérard*
BESANÇON, Musée des Beaux-Arts. Bistre wash on charcoal sketch, white paper 18·6 × 13 cm.

Probably drawn in 1778. Collections: P. A. Pâris; Bibliothèque Municipale de Besançon.

46 *Woman Reading a Book*
WASHINGTON, National Gallery of Art (Jules S. Bache collection). c. 1775. Oil on canvas 82 × 65 cm.

The theme of women reading was a favourite one with Fragonard. This painting, mentioned in the 1776 du Barry sale, shows him to have been as much at ease with women in studious moods as with erotic subjects. The yellows and ochres highlighted by the white ruff and the beautifully modelled face in pure profile make this a most appealing work. Its quiet, contemplative aspect was admired by the Impressionists, especially Renoir.

48 *Drawings for 'Don Quixote'*

These drawings, which have never been properly studied, were to be a great source of inspiration to Daumier. The projected edition of *Don Quixote* was never published; nineteen of the drawings were formerly in the collection of the Baron Vivant Denon.

(*left*)
a *Don Quixote Standing with Drawn Sword*
LONDON, British Museum. Chalk and wash 41·6 × 28·2 cm.

Sancho Panza is crouching behind.

(*right*)
b *The Priest Ordering the Destruction of Don Quixote's Books*
LONDON, British Museum. Chalk and wash 41·6 × 28·2 cm.

(*opposite*)
47 *The Reader*
PARIS, Musée du Louvre, Cabinet des Dessins. Pencil drawing on sepia wash 28·2 × 20·7 cm.

The Goncourts wrote of this drawing that 'Never did Fragonard create a woman out of so little. She stands out entirely clear, soft-limbed, almost transparent against the firm dark ground of the drawing. . . .'

49 *A Visit to the Nurse*

WASHINGTON, National Gallery of Art (Samuel H. Kress collection). c. 1778. Oil on canvas
90 × 70 cm.

Mentioned in the Leroy de Senneville sale (1784) and sold again at the Constantin sale in 1816 for
seven francs, the painting is the subject of a preparatory sketch in a private collection in
Switzerland. The theme of the happy family recurs frequently in the artist's middle years. A young
husband and wife, with their two small children, have come to visit their new-born child at the
foster-mother's house. With this kind of domestic interior scene Fragonard came close to Greuze,
The Village Bride, for instance, while remaining entirely without the latter's moralizing overtones.
The mother is still the mondaine young coquette, not very different from the girl in *The Swing*
(Wallace Collection). Until the late eighteenth century it was still the practice among middle- and
upper-class families to farm out their children to foster-mothers, and it was only when the lessons
of Rousseau's *Emile* (1762) began to take effect that breast-feeding became fashionable again.

50 *The Cradle*
AMIENS, Musée de Picardie. c. 1777–79. Oil on canvas 46 × 55 cm.

Left to the Musée de Picardie by the Lavalard brothers in 1894, this picture is one of numerous versions painted by Fragonard on the theme of the cradle or the happy family, all of them closely related. L. Réau suggests that their common source is the novel *Miss Sara Th* . . . (1765) by J. F. de Saint-Lambert, well-known in his day as a pastoral poet. The theory seems plausible for the novel contains a scene in which a young farmer and his wife go to look at their fifth child in a cot, followed by their four others, the same number which can be seen in Fragonard's painting. *The Cradle* may also have been an attempt to rival Greuze on his own ground of domestic virtue and conjugal happiness.

51 *The Schoolmistress*

LONDON, Wallace Collection. after 1780. Oil on canvas 28 × 37 cm. Signed on the blackboard after the letters of the alphabet: Fragonard.

The education of children was a favourite subject with Fragonard in his early forties, and this picture is related to a number of similar compositions, such as *The Little Preacher* (collection of A. Veil-Picard), *Say Please* (lost, but engraved in 1783) and *Education Does All* (Plate 52). The little boy is thought to be Fragonard's son Alexandre, who re-appeared almost identical in *The Reading Lesson* (Wildenstein, cat. no. 165). The setting for this humble village school seems to be the crypt of a disused Romanesque church, perhaps in Provence.

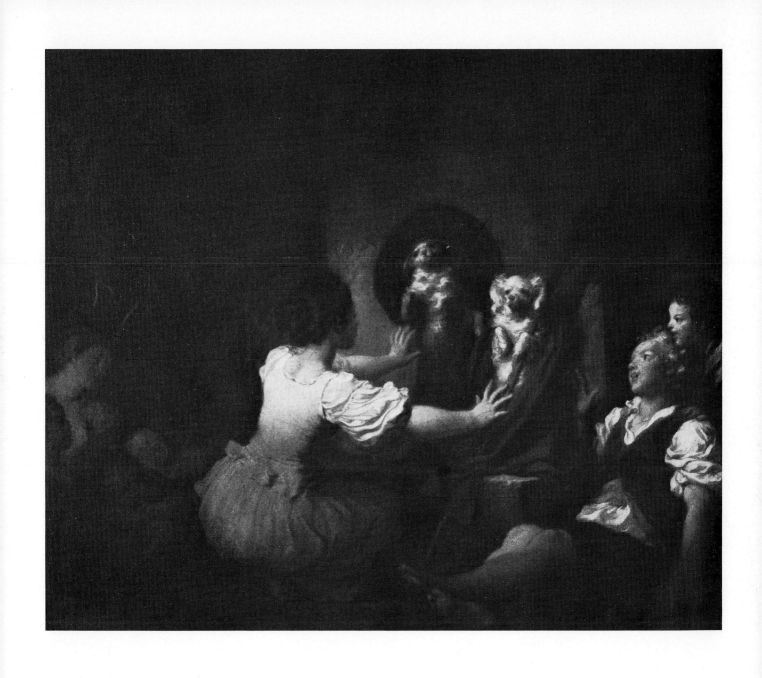

52 *Education Does All*
SÃO PAULO, Museum of Art. c. 1777–79. Oil on canvas 54 × 66 cm.

Engraved by N. de Launay as a pendant to *The Little Preacher*, this picture is a charming, if slightly sentimentalized version of *The Schoolmistress*.

53 *Boy Dressed as Pierrot*
LONDON, Wallace Collection. After 1785. Oil on canvas 60 × 50 cm.

In the Cope sale of 8 June 1872, where it was catalogued as by Boucher, this portrait was bought
at Christie's for Sir Richard Wallace. The extreme transparency of the surface of this picture
curiously anticipates Manet.

54 *Portrait of Madame Bergeret de Norinval*
PARIS, Musée Cognacq-Jay. c. 1780–89. Oil on canvas 60 × 50 cm.

This portrait is only attributed to Fragonard, but there can be no doubt about its superb quality. The handling of the paint is exceptionally light and delicate, and the woman's veil consists of no more than a gauzy white outline. The sitter's identity is uncertain. She may be a Mme. de Norenval, reader to Queen Marie Antoinette, or else Mme. Bergeret de Norinval, wife of Adelaïde-Etienne Bergeret, the nephew of Bergeret de Grancourt.

55 *The Stolen Kiss*
LENINGRAD, Hermitage Museum. c. 1786–89. Oil on canvas 45 × 55 cm.

The Stolen Kiss belonged to Stanislas Poniatowski, the last King of Poland, and was bought by
Tsar Nicholas II of Russia. It was transferred to the Hermitage in 1895. An engraving by
N. F. Regnault was published in the *Mercure de France* in 1788. The attribution to Fragonard is not
universally accepted; some see in this picture the hand of Marguerite Gérard. Others have pointed
out its affinity with the domestic scenes of Léopold Boilly. Several of Fragonard's late works,
including *The Bolt*, have the same curiously elongated forms, which may be a sign of his style
changing with the times; the sobriety of execution, the buff colours and dark green of the
background are unusual in Fragonard.

56 *The Bolt*
PARIS, Musée du Louvre. c. 1780. Oil on canvas 73 × 92 cm.

Originally in the collection of the marquis de Véri, *The Bolt* was acquired by the Louvre in 1972. There is another, sketchier version in the New York collection of the baroness von Cramm, but the Louvre painting is almost certainly the original from which Blot made his engraving of 1784. The presence of the apple on the table has been interpreted, perhaps over-ingeniously, as a symbol of the Fall, but there is little doubt about what is really happening in this picture. This is a straightforward seduction scene, treated in the rather lurid manner of an erotic novel. *The Bolt* suggests that, towards the end of his career, Fragonard turned away from the careless rapid strokes of his earlier style in favour of a more highly finished, enamelled surface.

57 *Drawings for La Fontaine*

Fragonard probably executed five series of drawings for La Fontaine's *Contes* which were never published. In 1795 an edition by Didot was inaugurated, in collaboration with A. de Saint-Aubin and Tillard, but the project was interrupted by the Revolution.

(left)
a *Le Paysan et son Seigneur*
BESANÇON, Musée des Beaux-Arts. Pen and ink on bistre wash 19·4 × 13·8 cm.

Collections: P. A. Pâris; Bibliothèque Municipale de Besançon.

(right)
b *Les Cordeliers de Catalogne*
BESANÇON, Musée des Beaux-Arts. Pen and ink on bistre wash 20·2 × 14·1 cm.

Collections: P. A. Pâris; Bibliothèque Municipale de Besançon.

58 *The Fountain of Love*
LONDON, Wallace Collection. c. 1780–88. Oil on canvas 64 × 56 cm. Signed at the bottom on the right: Fragonard.

Numerous copies exist of this work, which was engraved by N. F. Regnault as a pendant to *The Dream of Love* (Paris, Louvre) in 1785. The two figures are drawn as if in a vortex to the fountain, the source of love, the motive force in so much of Fragonard's work. The allegorical treatment of the theme and strange, rhythmical figures are more in the spirit of Prudhon and even Fuseli than the Rococo.

Bibliography

Fragonard's life and work:

Ananoff, A., *L'Oeuvre dessiné de Fragonard*, 3 vols. Paris, 1961–68.

Ananoff, A., 'Différents séries de dessins exécutés par Fragonard pour les Contes de La Fontaine', *Bulletin de la Société de l'Histoire de l'Art Français*, 1960.

Biebel, F. M., 'Fragonard et Madame du Barry', *Gazette des Beaux-Arts*, 1960.

Fragonard. Drawings for Ariosto (with essays by E. Mongan, P. Hofer and J. Seznec). London, 1946.

Guimbaud, L., *Saint-Non et Fragonard*. Paris, 1928.

Portalis, Baron R., *Honoré Fragonard, sa vie et son oeuvre*. Paris, 1889.

Réau, L., *Fragonard, sa vie et son oeuvre*. Brussels, 1956.

Thuillier, J., *Fragonard*. Geneva, 1967.

Wildenstein, G., *The Paintings of Fragonard*. London, 1960.

Wildenstein, G., 'La Fête de Saint-Cloud et Fragonard', *Gazette des Beaux-Arts*, 1960.

Wilhelm, J., 'Fragonard as a Painter of Realistic Landscapes', *Art Quarterly*, 1948.

General material:

Bachaumont, L. P. de, *Mémoires secrets*. London edition, 1777.

Bergeret de Grancourt, J. O., *Voyage d'Italie 1773–74, avec les dessins de Fragonard* (ed. J. Wilhelm). Paris, 1948.

Besançon: Inventaire général des dessins des Musées de Province. Collection Pierre-Adrien Pâris (ed. M. L. Cornillot). Paris, 1957.

De Brosses, C., *Lettres familières écrites d'Italie en 1739 et 1740*. Paris, 1869.

Diderot, D., *Salons* (eds. J. Adhémar and J. Seznec). Oxford, 1957–67.

Florisoone, M., *Le Dix-huitième siècle*. Paris, 1948.

France in the Eighteenth Century (catalogue of the exhibition at the Royal Academy of Arts). London, 1968.

Goncourt, E. and J., *L'Art du dix-huitième siècle*. Paris, 1859–75.

Levey, M. and Kalnein, Graf W., *Art and Architecture of the Eighteenth Century in France*. Harmondsworth, 1972.

Mercier, L. S., *Tableau de Paris*. 1781.

Mornet, D., *Le Sentiment de la nature en France au dix-huitième siècle*. Paris, 1907.

Portalis, Baron R., 'La Collection Walferdin et ses Fragonard', *Gazette des Beaux-Arts*, 1880.

Seznec, J., 'Don Quixote and his French Illustrators', *Gazette des Beaux-Arts*, 1948.

Wildenstein, G., 'L'Abbé de Saint-Non artiste et mécène', *Gazette des Beaux-Arts*, 1959.

Wildenstein, G., 'Un Amateur de Boucher et de Fragonard, Jacques-Onésyme Bergeret (1715–85)', *Gazette des Beaux-Arts*, 1961.